Handbook of
PRACTICAL
SOLID-STATE
TROUBLESHOOTING

PRENTICE-HALL SERIES IN ELECTRONIC TECHNOLOGY

Dr. Irving L. Kosow, *editor*

Charles M. Thomson, Joseph J. Gershon and Joseph A. Labok
consulting editors

PRENTICE-HALL INTERNATIONAL, INC., *London*
PRENTICE-HALL OF AUSTRALIA, PTY. LTD., *Sydney*
PRENTICE-HALL OF CANADA LTD., *Toronto*
PRENTICE-HALL OF INDIA PRIVATE LIMITED, *New Delhi*
PRENTICE-HALL OF JAPAN, INC., *Tokyo*

Handbook of
PRACTICAL
SOLID-STATE
TROUBLESHOOTING

JOHN D. LENK

Prentice-Hall, Inc., Englewood Cliffs, N. J.

This book is dedicated to my wife Irene
whose encouragement has made the book possible.

18   17   16   15   14   13   12

ISBN: 0-13-380642-1
Library of Congress Catalog Card Number: 75-167631
Printed in the United States of America

PREFACE

The practical techniques for troubleshooting solid-state equipment are quite different from those where vacuum tubes are used. Likewise, the troubleshooting techniques for digital circuits are different from those of linear circuits.

While it is true that the time-proven, basic methods of electronic troubleshooting (determining symptoms, localizing faulty circuits, isolating components, etc.) are good for all types of electronic devices, a series of *practical* troubleshooting techniques can be applied to solid-state and digital circuits.

Old-time radio and TV servicemen rely heavily on tube replacement as a troubleshooting method. It is generally assumed that most radio and TV troubles are the result of tube failure, and can be corrected by replacement. This assumption is definitely not true for solid-state equipment. Transistors are difficult to replace. Integrated circuits are even more difficult. Trouble must be isolated to a specific component or module *before* any attempt at replacement can be made.

Likewise, the logic circuits used in digital equipment (computers, numerical control, telemetry, etc.) are simple—when analyzed by themselves. But when hundreds or thousands of circuits are interconnected, as they are in the simplest of computers, the massive network can be an "electronic nightmare" unless a specific troubleshooting approach is used.

The purpose of this book is to fill the gap between the theory of solid-state and digital circuits, and the practical how-to of troubleshooting. The book is not just another review of semiconductor theory nor an over-simplified introduction to theoretical troubleshooting, but a technician-level book on how to pinpoint circuit and component faults. The book is written primarily for working technicians and field service engineers, but will prove an invaluable guide to the student technician who is about to face the realistic world of everyday troubleshooting.

v

Chapter 1 is devoted to basic troubleshooting techniques (evaluation of symptoms, fault isolation, etc.), with special emphasis on solid-state and digital circuits. Chapter 2 describes the use of test equipment and handtools associated with solid-state troubleshooting. This is particularly important since solid-state circuit faults are best located by analyzing test results (oscilloscope waveforms, voltage and resistance measurements, response to input signals, etc.). Illogical transistor circuit voltages, or abnormal waveforms, can often pinpoint solid-state circuit problems. Chapter 2 also covers construction of solid-state and integrated circuit (IC) equipment, as it relates to special troubleshooting and service problems (removal of IC packages, why IC equipment is serviced on a "system" basis, etc.). This discussion solves one of the problems faced by many technicians—the high density of solid-state construction, and the modular concepts of digital equipment.

Chapters 3 and 4 are devoted entirely to digital equipment. Chapter 3 introduces the reader to digital circuits, describing how to read logic diagrams. Chapter 4 covers those troubleshooting techniques that apply specifically to digital circuits.

Chapter 5 covers troubleshooting for basic circuits that often appear in laboratory and industrial equipments. Chapter 6 is devoted entirely to solid-state television.

It is assumed that the reader is already familiar with solid-state basics at a level found in the author's *Practical Semiconductor Databook* (Prentice-Hall, Inc., 1970). It is further assumed that the reader has a knowledge of basic test equipment at a level found in the author's *Handbook of Electronic Test Equipment* (Prentice-Hall, Inc., 1971). However, no direct reference to either of these books is necessary to understand and use the *Handbook of Practical Solid-State Troubleshooting*.

The author has received much help from a number of manufacturers and instructors. He wishes to thank them all, and is especially indebted to those at Hewlett-Packard, Textronix, Sencore, and Xerox Data Systems for their very special help. The author also strongly recommends *Electronic Troubleshooting,* prepared by the staff of Philco Technical Institute and published by Prentice-Hall, as an excellent introduction to theoretical troubleshooting.

J.D.L.

CONTENTS

1. BASIC TROUBLESHOOTING TECHNIQUES   1

1-1 Basic Troubleshooting Sequence 3
1-2 Failure Symptom Analysis 6
1-3 Localizing Trouble to a Module 12
1-4 Isolating Trouble to a Circuit 24
1-5 Locating a Specific Trouble 40

2. BASIC SOLID-STATE TROUBLESHOOTING   53

2-1 Measuring Transistor Voltages In-Circuit 53
2-2 Troubleshooting with Transistor Voltages 56
2-3 Universal Transistor Trouble Charts 58
2-4 Testing Transistors In Circuit (Forward Bias Method) 60
2-5 Transistor Testers 62
2-6 Testing Transistors Out of Circuit 64
2-7 Testing Diodes Out of Circuit 66
2-8 Testing Miscellaneous Solid-State Components 67
2-9 Testing Integrated Circuits 68
2-10 Working with Tools in Solid-State Equipment 70
2-11 Solid-State Servicing Notes 85
2-12 Solid-State Oscillator Bias Problems 87
2-13 Effects of Leakage on Amplifier Gain 91
2-14 Capacitors in Solid-State Circuits 93
2-15 Effects of Low Voltages on Resistance and Cold Solder Joints 96

vii

3. INTRODUCTION TO DIGITAL CIRCUITS   98

3–1 Logic Symbology.. 98
3–2 Binary Logic 99
3–3 Binary Numbers 100
3–4 Binary Coded Decimal System 100
3–5 Basic Digital Logic Elements and Symbols 102
3–6 Modification of Logic Symbols 104
3–7 AND Gates 106
3–8 OR Gates 108
3–9 NAND Gates...110
3–10 NOR Gates 111
3–11 EXCLUSIVE OR Gates 112
3–12 ENCODE Gates 113
3–13 Amplifiers, Inverters and Phase Splitters 113
3–14 Flip-Flops 115
3–15 Multivibrators 122
3–16 Delay Elements 124
3–17 Logic Symbol Identification 125
3–18 Logic Equations 130
3–19 Digital Component applications..................................131
3–20 Counter/Readout and Divider Circuit operation..................131
3–21 Conversion between Analog and Digital Information 137
3–22 Storage Registers 144
3–23 Shift Register 145
3–24 Sign Comparators 149
3–25 Adders 150
3–26 Comparison Circuit 152
3–27 Multiple Line Encoders and Decoders 153

4. BASIC DIGITAL TROUBLESHOOTING   156

4–1 Digital Test Equipment 157
4–2 Basic Digital Measurements 165
4–3 Digital Equipment Service Literature 183
4–4 Examples of Digital Equipment Troubleshooting 192

5. TROUBLESHOOTING LABORATORY AND INDUSTRIAL EOUIPMENT   199

5–1 Basic Feedback Amplifiers 200
5–2 Basic Operational Amplifiers 205
5–3 Basic Integrator (with Operational Amplifier) 223
5–4 Basic Differential Amplifier......................................226

5–5 Basic Power Supplies 231
5–6 Wien Bridge Circuits 235

6. TROUBLESHOOTING TELEVISION RECEIVERS   238

6–1 Low-Voltage Power Supply..238
6–2 High-Voltage Supply and Horizontal Output 245
6–3 Horizontal Oscillator and Driver 256
6–4 Vertical Sweep Circuits 265
6–5 Sync Separator Circuits 273
6–6 RF Tuner Circuits 278
6–7 Intermediate Frequency and Video Detector Circuits 286
6–8 Video Amplifier and Picture Tube Circuits 291
6–9 AGC Circuits 300
6–10 Sound IF and Audio Circuits 302

1. BASIC TROUBLESHOOTING TECHNIQUES

This chapter provides a summary or refresher of basic troubleshooting techniques for all types of electronic equipment. No matter what circuit is involved, a logical approach is needed to find and correct any fault. Certain troubleshooting requirements are imposed by complex and simple equipment, whether it be solid-state, digital, integrated circuit, or whatever.

For example, with any electronic equipment or system, you must know how the equipment works under *normal conditions*. This means that you must study the equipment to find out how each circuit works when operating normally. *Studying the equipment* can mean several different things. If the equipment is military or industrial, there will be an instruction manual that provides all sorts of information (theory of operation, operating and test procedures, performance specifications, parts lists and location diagrams, etc.). If the equipment is of the entertainment type (TV, radio, stereo, etc.), data sheets or fact sheets are usually available. While these data sheets do not include the elaborate descriptions found in technical manuals, they do contain condensed data (schematics, waveforms, voltage/resistance data, and test procedures) which are adequate for the type of equipment. In the absence of a data sheet, most home entertainment equipment is provided with a schematic diagram that also shows some waveforms and voltage/resistance data, as applicable.

No matter what data is available, study it thoroughly *before* attempting to troubleshoot the equipment. (In rare cases, there will be no data to study. If you have ever serviced equipment under these conditions, you will realize the value of technical data and of studying it thoroughly!)

You must know the function of *all controls and adjustments* and *how to*

1

operate them. Often, conditions that appear to be serious defects are the result of *operator trouble.* Also, as equipment ages, some adjustments may become critical. In any event, the technician must be capable of operating the equipment to do a good service job. Of course, it is not expected that a technician be an expert programmer to service a computer (although it would help). But the technician must be able to set the controls for the normal sequence of operation, and observe equipment performance for proper operation. Again, with complex equipment, the operating procedures are found in manuals or on data sheets. With simple equipment, the operating, procedure is obvious (hopefully) or is standardized (such as the operating procedures for TV, stereos, etc.).

You must know *how to use test equipment* and perform test/adjustment procedures. You will be hopelessly lost in any troubleshooting problem if you cannot use electronic test equipment effectively. Some equipment requires special test instruments (such as stereo generators for stereo receiver checks, spectrum analyzers for microwave tests, etc.). However, most troubleshooting procedures can be performed with three basic instruments: the meter (VOM and/or electronic meter), the oscilloscope, and a signal source (radio-frequency generator, audio generator, pulse generator, etc.). If you are not already familiar with these instruments, you would do well to make a thorough study of test equipment and its applications. (Such information is available in the author's *Handbook of Oscilloscopes: Theory and Application,* Prentice-Hall, Inc., Englewood Cliffs, N. J., 1968; *Handbook of Electronic Test Equipment,* Prentice-Hall, Inc., Englewood Cliffs, N. J., 1971; and *Handbook of Practical Electronic Tests and Measurements,* Prentice-Hall, Inc., Englewood Cliffs, N. J., 1969.)

You must know *how to use tools* to repair a trouble once it has been located. Most repairs can be made with basic tools (soldering tools, pliers, screwdrivers, etc.). However, special techniques must be used for certain equipment and circuits. Repair of printed circuit boards and removal and replacement of IC modules are typical examples. These procedures, and certain other special repair techniques are described in later chapters.

You must be able to *perform a checkout procedure on the repaired equipment.* No matter how simple the repair, this step should not be omitted. Sometimes, a trouble is the result of another trouble. If both troubles are not cured, the problem will still be there. A classic example of this is an intermittent short or arc between two points, which burns out a resistor on a printed circuit board or burns out an entire IC module. To make an adequate checkout, it is again necessary to understand the equipment and its operating procedure (hopefully found in the technical manual or data sheet). Generally, if equipment performs all of its operating functions in the proper sequence, it can be considered as repaired and ready for use. In some special cases, you may have to rely on the operator to make the check, and then report on the equipment's performance (such as with a complex computer or similar equipment). A checkout of the equipment after repair may point out the need for readjustment of controls.

Finally, you must be able to *analyze logically* the information of malfunctioning equipment, and *apply a systematic, logical procedure* to find the problem. In short, you must be able to think! The purpose of this book is to provide you with a logical and systematic approach to troubleshooting for any electronic equipment, as well as special procedures for specific types of equipment. However, the book cannot make you think.

1–1. BASIC TROUBLESHOOTING SEQUENCE

There are four steps in the basic troubleshooting sequence: (1) analyze the symptoms of failure, (2) localize the trouble to a complete functional unit or module, (3) isolate the trouble to a circuit within the module, and (4) locate the specific trouble.

On very simple equipment, or on equipment where there is *only one* functional unit, step 2 can be omitted.

1–1.1. Analyzing Symptoms of Failure

The first step in analyzing a failure symptom is to ask yourself if *you understand the symptoms.* Do you recognize abnormal operation, or the absence of a normal function? Do you know what the equipment is supposed to do? If the answers to these questions are "no," read the instruction manual and/or discuss the symptom with the operator. Of course, this must be modified to meet the situation.

For example, if you are repairing radios and a customer complains that "all of a sudden it started to sound funny," the obvious symptom is distortion. You can check this quickly by turning on the radio and tuning in a station. On the other hand, if you are servicing a digital device and the operator says "the readout skips all counts between 003 and 007," the symptom may not be obvious. Unless you service this type of equipment on a regular basis and have some magic method of pinpointing trouble, *read the manual and verify the symptom* yourself, using the correct operating procedure. If it does nothing else, this will eliminate the problem of operator trouble.

1–1.2. Localizing Trouble to a Module

Once the symptoms have been confirmed and analyzed, the next step is to localize the trouble to the most likely functioning unit of the equipment. The term *functioning unit* is used here to mean an electronic operation performed in a specific area of the equipment. For example, in a communications transceiver (such as a mobile citizens' band unit), the functioning units consist of a transmitter, receiver, power supply, and antenna-switching network. Each of these units may be mounted on a separate chassis or module, and all are interconnected by wiring on a common chassis. Equally possible is an arrangement where all of the functioning units are on one chassis. In more elaborate equipment, the modules may be physically separate and interconnected by cabling. In other cases, each functioning unit is mounted on printed circuit cards or boards that plug into a common chassis.

No matter what configuration is used, you must have a knowledge of *how each functioning unit operates* to localize the trouble in a systematic and logical fashion. With this knowledge, you can correlate the trouble symptoms previously observed. In effect, you can point to the most likely trouble area.

Using the example of a communications transceiver, failure to transmit (with normal reception) would most likely be the result of a failure in the transmitter unit or module. The antenna-switching network would be the second most likely cause of trouble.

If the symptoms are reversed (normal transmission, no reception), the receiver module is first on the list for detailed check. If transmission and reception are both absent (or abnormal), either the power supply or the antenna-switching network would be a logical starting point.

Keep in mind that localization of trouble to a module is not always sure. Sometimes, failure in one unit will show up as an abnormal symptom in another unit. Likewise, the operation of functioning units is often interrelated.

For this reason, a *functional block diagram* is provided for many complex equipment systems, as is discussed in a later section.

1–1.3. Isolating Trouble to a Circuit

After the trouble is localized (by analysis) to a single functioning unit, or if there is only one unit, the next step is to isolate the trouble to a specific circuit within the unit. If the unit has a *signal path* (such as a transmitter, receiver, amplifier, etc.), signals can be traced along these paths (with an oscilloscope or meter) or signals can be injected into the paths to isolate the problem. For example, if a transmitter module consists of an oscillator/multiplier/amplifier combination, and the oscillator and multiplier both produce a normal output (as measured on an oscilloscope or meter with an RF probe), but there is no output from the transmitter module, the amplifier circuit is the most likely cause of trouble.

Some equipment is provided with built-in indicators and test points. These should be used to full advantage wherever possible. In other equipment, there are no test points as such, but there are logical points between circuits where signals can be traced (injected or measured). For example, the base of a transistor is a logical input point for the circuit, while the collector and/or emitter are logical output points for a solid-state circuit.

Of course, it is not always possible to isolate trouble to a specific circuit by signal substitution and/or signal tracing. A power-supply circuit is a good example of this, since a power supply does not have signal paths as such. In these circuits, it is necessary to isolate trouble by means of voltage measurement, that is, by making a check for normal (and abnormal) voltages in the circuit.

1–1.4. Locating the Specific Trouble Within a Circuit

Once a trouble has definitely been traced to a circuit, the final step in troubleshooting (before actual repair and checkout) is to locate the defective component (or components). This does not mean that you must remove and test each component in the circuit (although this may be required in extreme cases). Instead, locating a defective component involves making a thorough *visual inspection* of the circuit, using the senses of sight, smell, hearing, touch, and, hopefully, common sense. Look for charred or burned components, smoke, arcing, and overheated parts. If you cannot see the trouble, the next step is to check the waveforms against those in the instruction manual or data sheet using an oscilloscope. Of course, not all circuits have waveforms. Even with those circuits that have waveforms and signal paths, you must ultimately rely on *voltage and/or resistance measurements*.

With solid-state equipment, voltage measurements are generally more effective than resistance measurements. The main reason for this is that the elements of a transistor will pass current if a voltage of the proper polarity is applied. For example, if an ohmmeter is used to measure an emitter resistor, and one end of the resistor is returned to the base (even through another resistance and a common ground), the base-emitter junction of the transistor could become forward-biased. This will cause current flow through the junction, with a *parallel resistance* around the emitter resistor being measured.

For this reason, much of the troubleshooting information found in later chapters is based on measurement of voltages at the elements of the transistor. (Of course, this is in addition to measurement of waveforms where applicable.)

1–1.5. Summary of Basic Troubleshooting Sequence

The basic troubleshooting sequence is summarized in Fig. 1–1. A detailed discussion of each step is made in the following sections of this chapter.

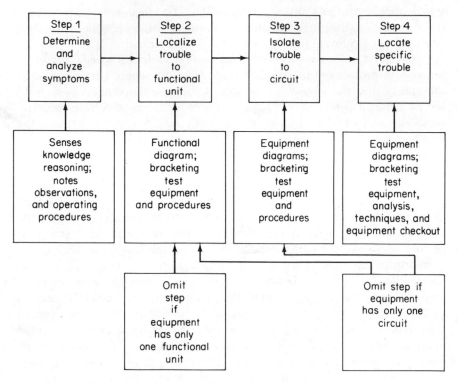

Figure 1–1. Summary of basic troubleshooting sequence.

1–2. FAILURE SYMPTOM ANALYSIS

One approach to failure symptom analysis is to observe the trouble symptom and then ask (yourself or the operator) questions regarding the symptom. In the case of total equipment failure, the questions and answers are often obvious. For example, if a television set fails to turn on, the questions would be "Is there power available? Is the power cord plugged in?

If the same TV set turns on but the picture is dull, the questions become a little less obvious. Typically, you could ask "Is the brightness control turned full on? Was the picture ever brighter than this?" If the answer is "yes" to both of these questions, you have definitely established *failure* or, at least, *poor performance*.

The fact that you asked a question regarding a "brightness control" indicates that you have some knowledge of the equipment and its operating controls. Of course, everyone familiar with television knows of a brightness control. The same knowledge is not so common for specialized types of electronic equipment. That is why the technician *must know* how the equipment normally operates and how to operate the controls of the equipment.

It is impractical, if not impossible, to maintain electronic equipment in first-rate operating condition if you do not know when the equipment is operating properly.

This goes back to what was discussed in Sec. 1–1. A technician should study every bit of information available (technical manuals, data sheets, schematics, etc.) and discuss the failure or performance symptoms with the operator or customer, *before attempting any trouble analysis.* In highly specialized electronic equipment, computers, for example, it is necessary to have *factory training* to do a good job of troubleshooting.

1–2.1. Practical Versus Theoretical Symptom Analysis

There are two approaches to trouble analysis that almost always result in failure—the extreme theoretical and the extreme practical.

The practical approach is typified by the old-time radio repairman. His first step is to replace every tube in the set, one by one, until the problem is cleared. If this does not solve the problem, he replaces the filter capacitor. If the problem is still present, he puts the set on the shelf, tags it as a "tough dog," and waits until he has enough time to replace each and every part, if necessary.

The same approach is sometimes carried over into more complex solid-state equipment, where components are mounted on plug-in circuit boards or cards. The plug-in units can be replaced, one at a time, until the trouble is cleared. In fact, some technical manuals recommend this procedure. If the plug-in modules are sealed and are to be replaced as a unit, the approach has some merit. But even under these conditions, the practical, parts-replacement approach without any trouble analysis is a waste of time.

If you can analyze the trouble, based on your knowledge of how the equipment works and with a complete run through of the operating sequence, you can often pinpoint the module where the trouble occurs.

At the other extreme, *the theoretical approach* is typified by those who feel that all troubleshooting can be "done on paper." An example of this is an engineer who insisted that the only data required in a technical manual was a complete set of logic equations. He felt that if the technician knew the logic any problem could be solved. Obviously, it would be quite helpful if the technician also knew such facts as the location of the operating controls, the operating sequence, how to make internal adjustments, and so on.

To sum up, a good troubleshooter must have an effective combination of theory and practical experience.

1–2.2. Operating Sequence

The first step in operating equipment through its normal sequence is to identify the *operating controls,* that is, to identify which are operating controls and which are *adjustment controls.* In simple equipment, the operating controls are usually obvious or highly standardized (such as radio or TV receiver controls). In complex equipment, the operating controls may not be obvious, and it is necessary to consult the technical manuals. In complex military equipment, separate operators' manuals are published to cover all operating controls, their function, and the normal sequence of operation.

In some cases, operating and adjustment controls overlap. As an example, on some TV sets the horizontal and vertical controls are located at the front, perhaps behind a hinged door; on other sets, the same controls are at the rear, in a somewhat inaccessible location.

Operating controls are all those controls readily accessible to the operator. More specifically, operating controls are all external switches and controls connected through a shaft or other mechanical linkage to internal circuit components (variable resistors, inductors, or capacitors) which can be adjusted *without* going inside the equipment enclosure. Operating controls must be manipulated to supply power to the equipment, to tune or adjust the performance characteristics, or to select a particular type of performance. Operating controls produce a *change* in circuit conditions by direct variation of resistance, inductance, and/or capacitance. In turn, this causes an indirect change in the circuit current or voltage. When operating controls are manipulated, you see or hear changes on the front panel information displays (front panel meters, cathode-ray tubes, loudspeakers, etc.).

On the other hand, adjustment controls are internal (inside the enclosure and not readily accessible) and have no direct effect on operating sequence. For example, compare the difference between a *volume control* and a *gain control.*

A volume control is a variable resistor (potentiometer) having a shaft extending outside the equipment enclosure. Generally, manipulation of this control causes a resistance change in a circuit in the audio section of a receiver, thus varying the gain of this section and, in turn, varying the sound from the loudspeaker.

A gain control is also a variable resistor that varies the gain of a stage. However, a gain control is (usually) an internal control requiring screwdriver adjustment, even though both the volume and gain controls are identical parts in identical circuit locations.

Both operating and adjustment controls have considerable significance in troubleshooting. Obviously, you must be able to identify controls and operate them properly, to check out equipment before and after troubleshooting. As discussed in later chapters, you must also set operating and adjustment controls to certain positions when signal tracing waveforms or making voltage/resistance checks.

In some cases, improper control settings can *appear as equipment failure* or poor performance. Likewise, improper control settings can *produce* equipment failure. Every circuit component has definite maximum voltage and current limits. The components must be operated below these limits to prevent damage. In some equipment, front panel meters are used to monitor conditions in critical circuits. It is *very important* that the operating controls be adjusted so as not to exceed the limits for those specified areas, especially when investigating trouble symptoms. If not, improper use of operating controls can result in even more damage to already defective equipment. Again, always refer to equipment manuals when troubleshooting unfamiliar equipment, especially noting any *operating precautions*.

In addition to exceeding maximum ratings and manipulating controls in an improper manner, there are certain other precautions associated with specific types of equipment. For this reason, it is especially critical that you study operating precautions when troubleshooting unfamiliar types of equipment. For example, the intensity or brightness control of a cathode-ray tube (in an oscilloscope, radar set, or TV set) should never be set to produce an excessively bright spot on the screen. Such a spot may burn the screen coating and decrease the usefulness of the cathode-ray tube.

To sum up, a knowledge of the *circuit changes* that take place when you adjust a control will help you think ahead of each step and anticipate any damage which the adjustment might produce. In any event, never make an adjustment in haste (with the possible exception of cutting off the main power switch!).

Once the controls (operating and adjustment) have been identified, the next step is to set all controls at their *normal or safe position*. In some cases, this is full off. More likely, the normal position is in the midrange of the controls.

Obviously, you should not set controls to their maximum or full-on positions (unless the manual so states). This is almost certain to aggravate a defective equipment condition. On the other hand, setting all controls to minimum or off could cause trouble, especially if the controls are operated out of sequence. For example, if a control disconnects the output of a push-pull (solid-state) amplifier from its load while power is applied, the amplifier transistors can become overheated and burn out.

In addition to preventing trouble or damage, there is a logical reason for setting controls to the normal position in the correct sequence, as a beginning step in troubleshooting: *you can gain information which will further define the trouble symptom*. Also, if the trouble symptom is the result of a control setting error (operator trouble) or control malfunction, it will be possible to cure the trouble by proper adjustment or replacement of the control. If the controls are properly set but the symptom continues, it is possible that an operating control is responsible for the trouble symptom. (In this case, the trouble would have to fall in the area of component failure.) If a control is faulty, this may be immediately apparent (such as a bad or burned spot on a composition potentiometer). Of course, additional information may be required to find an electronic failure in a control, since the trouble symptom produced may also point to other electronic failures.

1–2.3. Symptom Recognition

Symptom recognition is the art of identifying the *normal* and *abnormal* signs of operation in electronic equipment. When the equipment is not performing properly, it will display some sign of abnormal performance, and when the equipment is performing properly, it will display signs of normal performance. You must be able to distinguish both.

For example, a normal TV picture is a clear, properly contrasted representation of an actual scene. The picture should be centered within the horizontal and vertical boundries of the screen. If the picture suddenly begins to "roll" vertically, you would recognize this as a trouble symptom because it does not correspond to the normal performance which is expected.

1–2.4. Equipment Failure Versus Poor Performance

Equipment failure means that either the entire piece of equipment or some part (functional unit) of it is not operating properly. For example, the total absence of sound from a radio receiver when all controls are in their proper position indicates complete or partial equipment failure.

Poor performance can be defined as equipment that is working but is presenting information (or an output) that does not correspond with design specifications. This performance may range from a nearly perfect operating condition to the condition of barely operating. For example, the presence of a hum in the sound from a radio receiver indicates poor performance.

Equipment failure is usually easier to recognize and possibly easier to pinpoint than poor performance.

1–2.5. Evaluating Symptoms

Symptom evaluation is the process of finding a *more detailed description* of the trouble symptom. The purpose of symptom evaluation is to help you understand fully what the symptoms are and what they *truly indicate,* to gain futher insight into the problem.

The technician who charges into the equipment, measuring waveforms and taking voltage/resistance readings, without studying the symptoms, is wasting time and energy. Even worse, such a procedure can easily lead the technician astray.

To evaluate anything, including electronic trouble symptoms, you should get as much information as possible. The mere recognition of an original trouble symptom does not in itself provide enough information to decide on the probable cause or causes of trouble. Keep in mind that many different faults can produce similar trouble symptoms. For example, in the case of receiver hum, such a symptom may be caused by poor filtering in the power supply, line voltage interference, external interference, or many other defects.

To evaluate a trouble symptom, you will have to manipulate the equipment controls associated with the symptom and apply your knowledge of basic electronics, supplemented with information gained from the equipment manuals.

However, the mere adjustment of the operating controls to their normal positions is not the complete procedure followed during symptom evaluation (although finding an incorrect control setting is part of the complete process). For example, recognizing that the screen of a TV set is not lighted is not sufficient information for you to decide exactly what could be causing the trouble. This symptom could mean that the cathode-ray tube filament is burned out, that there is some disorder in the internal circuitry associated with this tube (high-voltage power supply), that the brightness control is turned down too low, or even that the equipment is not turned on. Think of all the time you may waste if you tear down the equipment and begin testing, when all that is needed is to set the power switch to *on*, adjust the brightness control, or simply plug in the main power cord!

1–2.5.1 Data Recording

To evaluate trouble symptoms completely, all indications must be evaluated in relation to one another, as well as in relation to the overall operation of the equipment. One handy method of doing this is to record all of the data (control positions and indications) that are found.

Such data recording will permit you to study the information before jumping to conclusions. They will also permit you to check the equipment manual (or data sheet) and compare the recorded information with the written material. Finally, by recording all control positions and the associated data, you can quickly reproduce the information and check to see that it is correct, as well as to put the equipment in exactly the operating condition that you wish to test.

1–2.6. Performance Evaluation

The senses of sight and hearing allow you to recognize the symptoms of normal and abnormal equipment operation and thus to evaluate the performance of the equipment. This is because most electronic equipment yields some information that an operator or technician can either see or hear. For example, electrical information to be presented as sound must be applied to a loudspeaker or a headset. A visual display results when the electrical information is applied to a cathode-ray tube or to an indicating meter, which is built into the equipment control panel, and can be viewed by the operator. Pilot lights (and interpolation lights on computer-type equipment) also provide a visual indication of equipment operation.

In some equipment, the display of information may be the only job performed (such as a decade readout), or it may be a secondary job. Either way, knowledge of the *normal equipment displays* will help you recognize abnormal displays.

A final word on failure symptom analysis: *do not* tear into the equipment until *all of the symptoms* have been evaluated.

1–3. LOCALIZING TROUBLE TO A MODULE

On complex equipment, the second major step in troubleshooting is to localize the trouble to a module or functioning unit of the equipment. This step can be omitted when there is only one functioning unit. (A functional unit is a specific area or section of the equipment in which some *specific electronic operation* is performed, such as the RF section, audio section, power supply, etc). This section may or may not be physically separate from other sections, depending on the complexity and design of the equipment. When combined, the functional units make up an *equipment set,* such as a radar set (where there are physically separate functional units), or a transceiver (where the functional units are all mounted on a common chassis in a common enclosure).

To localize the trouble to a functional unit, you must make use of symptom recognition and evaluation to find which units are the *probable cause* of trouble. You then find which of the selected functional units is the source of trouble by using test equipment to check inputs and outputs in a logical, systematic manner.

After completing this step, you verify that you have pinpointed the functional unit with the trouble, making sure that it is the *only unit at fault.*

To localize trouble to a module or functional unit, you must evaulate symptoms and make a *mental decision* as to the most probable area in which the trouble may be. Up to this point, you *have not* torn into the equipment, made internal adjustments, or used test equipment other than the built-in indicators of the equipment (if any). Now you are going to start with test equipment, systematically checking each unit selected (by mental decision) until the actual faulty one is found. Usually, you will use *signal-tracing* and/or *signal-injecting* test equipment (such as meters with probes, oscilloscopes, and signal generators) in the localizing process. If none of the functional units on your list of selections performs improperly, you must make a *return path* and reevaluate the symptom. Possibly, you will have to get more information. On solid-state equipment where functional units are contained in plug-in modules, it may be practical to replace each functional unit, one at a time, until the faulty module is located (and the trouble is cleared). However, it is usually quicker to localize trouble by means of test equipment.

1–3.1. Relationship of Functional Divisions of Equipment to Troubleshooting Sequence

To properly understand the troubleshooting sequence, particularly the localizing step, it is necessary to understand the relationship of the four major troubleshooting steps to the makeup of electronic equipment.

Any electronic device is made up of individual *components* (or parts) which are connected together to form *circuits.* These circuits perform electronic subfunctions (except in very simple equipment where the circuit performs a complete function). The circuits are connected to form *functional units,* which perform major electronic functions. The functional units are interconnected to form a *system* or *set,* which performs overall operational functions. On less complex equipment, the functional unit (or module) performs the overall operational function. This is the case where there is only one functional unit.

Note that the terms *function* (an operational subdivision) and *unit and module* (physical subdivisions) are used interchangeably. Although this may seem confusing it is typical of the terms found in technical manuals and other troubleshooting literature.

Figure 1-2 clarifies the relationship of troubleshooting and equipment functions. Note that there are three configurations. In Fig. 1-2a the equipment consists of more than one functional unit, while Fig. 1-2b shows the relationship where there is only one functional unit. Figure 1-2 c shows the relationship in which one or more of the circuits are in IC form.

First, there is the *system* or *set* (radar set, computer system, etc.) designed to perform an overall operational function. Notice that the *analyze symptoms of failure* troubleshooting step is associated with this classification.

Next, the system or set is divided into *functional units* or *modules* (receiver, transmitter, modulator, etc.), each designed to perform a major electronic function vital to the overall operational function. The *localize* step is associated with this classification. (With only one functional unit, the localize step is omitted.)

The next division is the *circuit group* or individual *circuit.* The circuit group (RF section, IF section, converter, etc.) is a subdivision of the function-

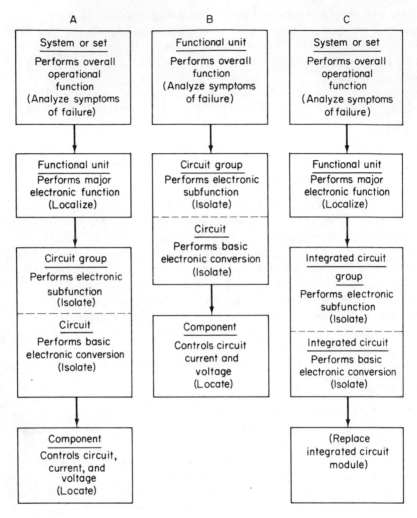

Figure 1–2. Relationship of troubleshooting sequence to equipment functions and divisions.

al unit and performs an electronic subfunction. The individual circuit (RF amplifier, IF amplifier, local oscillator, etc.) performs the basic electronic conversion process (input-conversion-output) within the circuit group. The *isolate* step of the troubleshooting sequence is concerned with determining which circuit is at fault.

The last classification is the component—the device (transistor, diode, resistor, etc.) that controls the circuit current and voltage—which is *located* in the last step of the troubleshooting sequence.

1–3.1.1. IC Troubleshooting Sequence

The troubleshooting sequence must be modified when ICs are used. The last step (locate defect to a part) is unnecessary (if not impossible) to accomplish. ICs are made up of many parts to form one circuit or several circuits. The ICs are sealed and are replaced as a package. Therefore, once trouble is located in the IC, the troubleshooting sequence is complete, and the repair/checkout phase starts. For example, if an IC is used to form the complete IF section of a TV set and trouble has definitely been isolated to the IF section (by means of signal tracing, voltage checks, or whatever), the next logical step is to replace the IC and check the set. Problems associated with troubleshooting ICs are discussed in Chapter 2.

1–3.2. Functional Block Diagrams

A functional block diagram is an overall representation of the functional units within the equipment, as well as of the *signal flow paths* between units. Figure 1–3 shows a typical functional block diagram for an amplitude-modulated transceiver set composed of six functional units. In some technical manuals, such an illustration is referred to as an *overall block diagram.* Keep in mind that the functional units may be physically separate, may be located on different sections of a single chassis, or may be separate plug-in modules mounted on a common chassis.

There is no indication in Fig. 1–3 as to how each function is accomplished. Thus, each unit may consist of a variety of circuits or stages, each performing its own major electronic function. For example, the transmitter unit may contain an RF oscillator circuit, an RF voltage amplifier circuit, and several RF power amplifier circuits. However, the major electronic function of the transmitter is to produce an RF carrier wave.

Notice that the connecting lines between the various functional blocks (or units) represent *important signal flow connections,* but that the diagram does not necessarily show where these connections can be found in the actual equipment circuitry. In some cases, additional information such as frequencies and input-output waveforms may be included to show how far each type of signal progresses through the equipment. The layout interpretation of the connecting lines and the representation of the blocks will depend upon the complexity of the equipment and the care with which the technical manual is written.

1–3.2.1. Interpreting the Functional Block Diagram

Although a functional block diagram does not show physical relationships (there is no relationship whatsoever between the physical

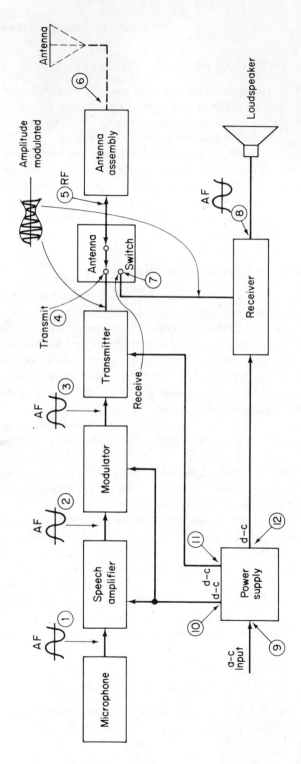

Figure 1–3. Functional (overall) block diagram of transceiver.

16

location of the units within the equipment and the block arrangement in the functional block diagram), the diagram does provide a general representation of the major functional units and the important *signal relationships between them*. Therefore, the block diagram can be used to understand the equipment's overall purpose (what major functions must be accomplished to perform the overall designed task?).

In most technical manuals, the overall block diagram is supplemented with text material describing the overall *theory of operation*. For example, if the diagram of Fig. 1–3 were used in a typical technical manual, the supporting theory of operation would be somewhat as follows.

During transmission, the *microphone* converts the sound information to be transmitted into an electrical signal of audio frequency. The *speech amplifier* amplifies the AF signal and applies it to the *modulator* which, in turn, applies the same signal to the transmitter in such a manner as to cause the amplitude of the RF carrier signal to vary at an audio rate.

In addition to providing the RF carrier signal, the *transmitter* amplifies the signal to a suitable level. The *antenna assembly* converts the RF signal into electromagnetic energy.

When the transceiver is serving as a receiver, the antenna assembly picks up the incoming RF signal and applies it to the *receiver unit*. The receiver removes any AF information from the signal and applies the AF portion of the signal to the loudspeaker.

The *power supply* converts the a-c line voltage into d-c voltages suitable for operation of the various units and circuits. The *antenna switch* connects the antenna to the transmitter output or receiver input, depending on the mode of operation (transmit or receive).

1–3.3. Good Input/Bad Output Technique

Although the functional block diagram is an essential reference for localizing the trouble to a functional unit, accomplishment of this step is based upon the good input/bad output technique, together with symptom analysis and/or signal-tracing tests.

The good input/bad output technique (sometimes known as *bracketing*) provides a means of narrowing the trouble area to a single faulty functioning unit, to a circuit group, and then to a faulty circuit.

The process involves test equipment (to measure inputs and outputs), mental evaluation of symptoms, or both. Either way, the process starts by placing brackets (at the good input and the bad output) on the block diagram. With the brackets properly positioned, you know that the trouble exists somewhere between the two brackets.

The technique is to move the brackets one at a time (either good input or bad output) and then make tests to find if the trouble is within the new bracketed area. This process continues until the brackets localize a single defective unit.

The most important factor in bracketing is to find where the brackets should be moved in the elimination process. This is determined from *your deductions based on your knowledge* of the equipment and the symptoms. All moves of the brackets should be aimed at localizing the trouble with a *minimum of tests*.

1–3.4. Examples of Localization Technique Using Functional Block Diagram

The good input/bad output technique can be used with or without actual measurement of signals at test points. That is, sometimes localization can be made on the basis of symptom evaluation alone. In practical troubleshooting, *both* symptom evaluation and tests must be made, often simultaneously. The following examples show how the technique is used in both cases.

Assume that a transceiver is being serviced, and that a functional block diagram (with corresponding test points) similar to Fig. 1–3 is provided in the technical manual. Note that test point 2 is at the output of the speech amplifier and at the input of the modulator, simultaneously. Likewise, test point 3 is at the output of the modulator and at the transmitter input.

Test points 10, 11, and 12 could be considered as inputs to the various functional units. However, these test points are at d-c voltages, rather than signal voltages, and will be considered as outputs from the power supply.

Also note that test points 1, 2, 3, and 8 are AF voltages, while test points 4, 5, 6, and 7 are RF voltages, amplitude modulated by AF voltages.

The voltages at all test points, except at 4, 5, 6, and 7, can be measured with a meter or oscilloscope, using a *basic probe* or test leads. The voltages at test points 4, 5, 6, and 7 will require an *RF probe* (to measure the RF carrier) and a *demodulator probe* (to measure the amplitude-modulated AF voltages).

Example No. 1. Assume that the operator reports no reception. You, as the technician, confirm the symptom by operating the equipment through its normal sequence. The receiver tuning controls (or channel selector) and the receiver volume controls have no effect on the symptom. When the operating controls are set to the receive position, there is no sound from the loudspeaker.

Without using any test equipment, it is possible to eliminate several functioning units based on the trouble symptom. Since the problem is in reception, the speech amplifier, modulator, and transmitter can be eliminated. However, the receiver, antenna assembly, antenna switch, and power supply could be at fault. (Using the bracketing system, you would place brackets around these four possibly faulty units.)

The next step is to eliminate three of the four units, by test or substitution. The trouble is then localized to the one remaining unit. The localization procedure could take one of two courses, depending upon design of the equipment.

If the functional units are plug-in modules, as is the case in much solid-state equipment, each of the four modules can be replaced in turn, until the trouble is cleared. For example, if replacement of the receiver module restored normal operation, the defect would have been in the receiver module. This could be confirmed by plugging the suspected defective module back into the equipment. (Although this confirmation process is not a necessary part of theoretical troubleshooting, it is wise to perform it from a practical stand-point. Often, a trouble symptom of this sort can be caused by the plug-in module making poor contact with the chassis connector or receptacle.)

If the functional units are not of the plug-in type or if they are not readily replaceable, localization must be made by means of tests. In the case of the power supply, the d-c voltage (at test point 12) to the receiver should be measured. If this d-c voltage is correct at the receiver, the power supply can be eliminated.

With the trouble narrowed down to the receiver, antenna switch, or antenna assembly, the next step is to decide whether *signal injection* or *signal tracing* will permit the fastest localization. (Both signal injection and signal tracing are discussed in greater detail in later sections.)

With signal injection, a signal generator is tuned to the operating frequency of the receiver (RF) and modulated by an AF tone (say, 1 kHz). The signal generator output is injected at test points 7, 5, and 6 (in that order), with the transceiver controls in the receive position of operation. If a tone is heard in the loudspeaker with a signal at test point 7, the receiver can be eliminated, and trouble is most likely in the antenna switch or antenna assembly. If a tone is heard with a signal at 5 but not at 6, the antenna assembly is the most likely cause of trouble (perhaps a broken antenna connector).

With signal tracing, a signal generator is tuned to the operating frequency of the receiver and is modulated by an AF tone. The signal generator output is connected to test point 6, and the signal is traced (by means of oscilloscope or meter with RF and demodulator probes) at points 6, 5, and 7 (in that order). If the signal is present at test point 6 but not at 5, the antenna assembly is at fault. If the signal is present at 5 but not at 7, the antenna switch is defective. If the signal is present at 7 but there is no tone from the loud-speaker, the trouble is localized to the receiver.

Example No. 2. Assume that the operator again reports no reception, which you confirm. However, now you find that there is a scratching sound from the loudspeaker when you adjust the volume control. You also check the transmitter and find that all front panel meters provide normal indi-cations.

In this case, the trouble localization process is similar to that described in Example No. 1, except that the power supply can be eliminated immediately, *in theory,* without further tests. (From a practical, troubleshooting viewpoint, the more units that can be eliminated from suspicion, without testing, the better.)

If the power supply were defective as a complete unit, the transmitter function would not be present nor would the panel meters provide normal indications. If the power supply were functioning partially (d-c voltages to all units but the receiver), the receiver would be "dead," and there would be no sound from the loudspeaker when the volume control is operated.

In practice, trouble localization should *begin with* signal tracing or signal injection at the receiver, antenna switch, and antenna assembly test points, as described. *Most likely* the trouble will be in one of these three units. But it is not absolutely certain that the power supply has been cleared of possible fault. Assume, for example, that the power supply is delivering correct voltages to all units except the receiver because of a high-resistance lead to the receiver. Or assume that the receiver voltage comes from a separate circuit within the power supply, and only that circuit is defective and is delivering a low voltage.

This point is brought out to show the difference between theoretical and practical troubleshooting. In theory, with the symptoms as described, the power supply is eliminated from any possibility as a defective unit. In practical troubleshooting, the same symptoms localize trouble to the receiver, antenna assembly, and antenna switch as the *most likely* areas of trouble and as the *starting point* for further trouble localization.

Example No. 3. Assume that reception is normal on all frequencies or channels of the transceiver, but there is no transmission on any channel. (You confirm that it is impossible to contact anyone on all channels, and you note that the transmitter tuning-meter indication is low but the modulator meter is correct.)

The *most likely defective* functioning units are the transmitter, antenna switch, and antenna assembly. If the speech amplifier or modulator is defective, the modulator meter will show an abnormal indication (or there will be no indication). If the receiver is defective, there will be no reception. If the power supply is defective, none of the functions will be correct. Of course, if the power supply is delivering a low voltage to the transmitter but correct voltages to all other units, the same symptoms occur. However, this is not a most likely condition.

If the transmitter, antenna switch and antenna assembly are of the plug-in type, they can be replaced in turn until the trouble is cleared. If the units are not readily replaceable, tests must be made to localize the trouble.

Since a transmitter provides or generates it own signal, *signal tracing* is the most logical method of test (in preference to signal injection). With signal tracing, the transceiver is operated in the *transmit* mode, and the signal is traced (by means of an oscilloscope or meter with RF and de-modulator probes) at points 4, 5, and 6 (in that order). If the signal is absent at test point 4, the transmitter unit is at fault. If the signal is present at 4 but not at 5, the antenna switch is defective. A good signal at 5 with no signal at 6 indicates a defective antenna assembly.

1–3.5. Which Unit to Test First

When you have localized the trouble to several functional units, you must decide which unit to test first. This is especially true when the units are not readily replaceable. Several factors should be considered in making this decision.

Generally, if you can make a test that eliminates several units, that test should be made before making a test that eliminates only one unit. This re-quires an examination of the block diagram and a knowledge of how the equipment operates. It also requires that you apply logic in making the decision.

The accessibility of test points is the next factor to consider. A test point could be a special jack located at an accessible spot on the equipment, such as the front panel or the chassis. The jack (or possibly a terminal) is electrical-ly connected (directly or by a switch) to some important operating voltage or signal path. A test point could also be any point where wires join or where parts are connected together.

Another important factor to consider is past experience and history of repeated failures. Past experience with similar equipment and related trouble symptoms, as well as the probability of unit failure based upon records or repeated failures, should have some bearing upon the choice of a first test point.

1–3.5.1 Example of Choosing the First Test Point

Assume that the transceiver of Fig. 1-3 is being serviced. There is no audible signal from the receiver, and there is no effect produced by manipulating the receiver volume control. These are the same symptoms described in Example No. 1. Trouble is localized to the power supply, re-ceiver, antenna switch, or antenna assembly. The test points involved are 5, 6, 7, 8, and 12. Assume that all of these tests points are equally accessible, that the receiver has a previous history of failure in the local oscillator section, and that it is decided to test by signal-injection and voltage measurement.

Test point 8 can be eliminated as a first test choice, since it will only prove the loudspeaker to be good or bad. Test point 6 can be eliminated. No response to a signal injected at 6 will prove nothing. This is the same as the basic symptom of no reception. Test point 5 is a poor first choice. No response to a signal injected at 5 will prove only that the antenna assembly is *probably good.*

Either test point 7 or 12 is a good choice. If there is a good response to a signal injected at 7, both the power supply and receiver are cleared, and trouble is localized to the antenna or switch. If response is bad with a signal at 7, the trouble is localized to the receiver or power supply. A check of the voltage at point 12 will then isolate the trouble to one unit (if the voltage at 12 is correct, the trouble is most likely in the receiver).

Now assume that all conditions are identical, except that the receiver shows some sign of operation (background noise from the loudspeaker that is adjustable by means of the volume control, but no signal).

In this case, the most logical first test point is 7 rather than 12. The fact that the volume control will adjust the noise level indicates that at least the audio stages are operating properly. Since these circuits require voltages from the power supply, it is safe to assume under these conditions that the power supply output is satisfactory. This fact, plus the history of previous local oscillator failure, leads you to conclude (tentatively) that the receiver is at fault.

Consider another example of selecting the first test point, based on evaluation of symptoms. Assume that the transceiver of Fig. 1-3 is being serviced. Now, reception is good on all channels, but transmission is erratic. The transmitter tuning-meter indication is low, and the modulator meter reading is erratic.

Trouble is localized to the power supply, speech amplifier, modulator, and transmitter. The test points involved are 1, 2, 3, 4, 10, and 11. All test points are equally accessible, there has been no previous history of equipment failure, and it is decided to test by voltage measurement and signal tracing (a logical approach for a transmitter section).

Test point 1 can be eliminated as a first test choice, since it will only prove the microphone to be good or bad. Test point 4 can be eliminated. An erratic signal at 4 will prove nothing. This is the same as the basic symptom of erratic transmission.

Test points 2, 10, and 11 are better (but still poor) choices. An erratic signal at 2 will prove a defect only in the speech amplifier or in one section of the power supply. A bad voltage reading at 10 or 11 isolates trouble to one section of the power supply.

Test point 3 is the best choice for a first test. If the signal is erratic, trouble is localized to the speech amplifier, modulator, or one section of the power supply. A good signal at 3 will localize the trouble to the other power supply section or the transmitter. *Always aim for simultaneous elimination of several functional units, whenever accessibility of test points permits.*

Now assume that an erratic signal is found at point 3. The next most logical test point is 10. If the voltage is abnormal, the problem is in the power supply. If the voltage is normal, the trouble is in the speech amplifier or modulator (a further test of the signal at 2 will confirm which unit is at fault).

1–3.6. Modifying the Troubleshooting Procedure

Anyone who has had any practical experience in troubleshooting knows that all of the steps in a localization sequence can rarely proceed in a textbook fashion. In some cases it may be necessary to modify your troubleshooting procedure as far as localizing the trouble is concerned. Experience with particular equipment will provide special knowledge which may simplify localizing the trouble. The physical arrangement or configuration of a system may pose special troubleshooting problems.

For example, assume that the transceiver of Fig. 1-3 is being serviced. Transmission and reception are weak on all channels. The modulation meter indication is normal, but the transmitter meter shows a low indication. All stations can be received, and the volume control works, but all signals are weak.

With these symptoms, either the power supply or the antenna assembly may be at fault. Both of these units are common to *transmission and reception*. The antenna assembly is a more logical selection than the power supply (since the modulator meter shows a normal indication). However, there may be a practical problem to consider.

Although the antenna assembly is a logical first choice, it may be advisable to test the power supply unit first if the antenna assembly is not readily accessible, or if tests on it are more difficult to make. (Usually, an antenna assembly for a transceiver consists of an antenna, lead-in, feed-through, connector, and possibly a loading network.)

Now suppose that both the antenna assembly and the power supply check out satisfactorily (unlike what you expected or were led to believe by the data in the transceiver's technical manual). Somewhere in making your selection you may have overlooked some data in the symptom evaluation, or you may have missed a possible faulty unit.

Under these circumstances, your first thought might be to reconsider the original symptom, or possibly to return immediately to the symptom evaluation phase of localization. You would do better to retrace your steps one at a time, rather than jumping over the whole group of steps back to the starting point. First assure yourself that, from your knowledge of the symptoms and the symptom evaluation, you have *chosen all possible faulty functional units.* Of course, if you cannot logically include other functional units in your list of selections, you should go back even further to the symptom evaluation phase to see if you have overlooked anything.

1–3.7. Verifying Trouble

If the equipment design is such that the functional units are plug-in, or readily replaceable, the trouble can be verified easily. If you replace a functional unit and the trouble is cleared, this simultaneously verifies that the replaced unit was at fault.

When it is not easy to replace a unit or no replacement is available and you must go inside the unit to locate the defective circuit or part, you may save considerable time and effort if you verify the trouble, that is, reconsider the fault in this unit that could logically produce the trouble symptom. Also, does the fault fit the associated information found during symptom evaluation?

For example, if the speech amplifier in Fig. 1-3 is defective (no output), the transmitted signal will be a constant-amplitude RF signal, and the receiver will function normally. If these are the *only* symptoms, the choice of the speech amplifier is a good one.

The AF signal from the speech amplifier unit is eventually used to amplitude modulate an RF carrier signal in the transmitter. The absence of this AF signal means that no modulation will occur. Yet, the carrier signal that is produced in the transmitter will still be generated and will appear as a constant-amplitude (zero-modulation) RF signal. The receiver is not associated with the AF signal from the speech amplifier.

1–4. ISOLATING TROUBLE TO A CIRCUIT

The first two steps (symptoms and localization) of the troubleshooting procedure give you the initial symptom information about the trouble and describe the method of localizing the trouble to a *probable faulty* functioning unit. Both steps involve a minimum of testing, except by operation of the equipment controls.

In the *isolate* step, you will do extensive testing (with test equipment) in order to isolate the trouble—first to a group of circuits within the functional unit, and then to the specific faulty circuit. In the case of integrated circuits, where one IC is used to form a group of circuits within a functional unit, the trouble is isolated to the IC unit input. No further isolation is necessary since parts within the IC cannot be replaced on an individual basis. This same condition is true of some solid-state equipment where groups of circuits are mounted on sealed, replaceable boards or cards.

No matter what physical arrangement is used, the isolation process follows the same reasoning you have used previously: the continuous narrowing down of the trouble area by making logical decisions and performing logical tests. Such a process reduces the number of tests which must be performed, thereby saving time and reducing the possibility of error.

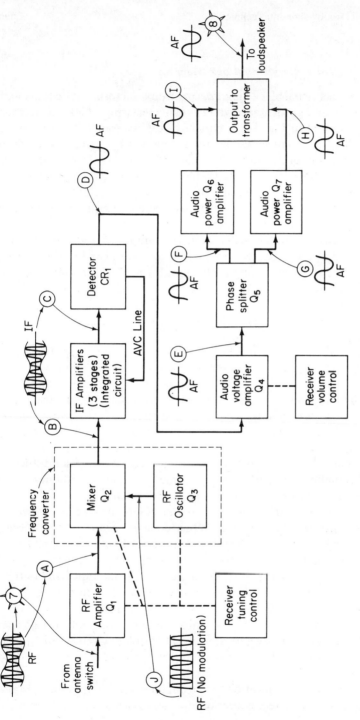

Figure 1–4. Servicing block diagram of receiver.

1–4.1. Servicing Block Diagrams

Servicing block diagrams provide a pictorial guide for isolating the trouble. Although there are variations in servicing block diagrams, the diagram illustrated in Fig. 1-4 is typical of the diagrams found in military and well-prepared commercial technical manuals. Figure 1-4 represents the receiver portion of the transceiver shown in Fig. 1-3. In a good technical manual there will also be a servicing block diagram for every other unit in the transceiver set. On simple equipment, such as those covered only by a data sheet rather than a technical manual, the entire equipment will be represented by one servicing block diagram.

In using the servicing block diagram, the symptoms and related information obtained in the previous steps should not be discarded now or at any time during the troubleshooting procedure. From this information you can identify those circuit groups that are probable trouble sources.

1–4.1.1 Interpreting Servicing Block Diagrams

The servicing block diagram can be seen in Fig. 1-4. Keep in mind that this diagram is representative of those in military technical manuals.

Notice first that all circuits within the functional unit are enclosed by a dashed line, as are circuits comprising the circuit groups within the unit. Within each dashed line is the name of the functional unit or circuit group it represents.

Main signal or data flow paths are represented by heavy solid lines, and secondary signal or data paths are represented by lighter solid lines. Arrows on the lines indicate the direction of signal or data flow.

Operating controls are connected to the circuit or circuits which they control by dotted or dashed lines. When depicted properly, the operating control will be labeled on the diagram. This name should correspond exactly with the control name that appears on the front panel.

Waveforms are given at several points, usually at the input and output of each circuit. Test points are identified by both letter and number. Generally, numbered test points (often starred as shown) represent points which are useful in localizing faulty functional units. Lettered test points (often circled as shown) represent points which are helpful in isolating faulty circuit groups or individual circuits. (This letter/number combination of test points follows the military style.)

Note that the physical location of the circuit groups within the equipment has *no relation* to their representation on the servicing block diagram.

1–4.1.2 Circuit Groups Versus Individual Circuits

It is important that you recognize *circuit groups* as well as individual circuits. (A circuit group is a subdivision of the functional unit that performs a *single electronic subfunction*. In some cases, a subfunction is performed by individual circuit.)

If you can subdivide a functional unit into circuit groups, you can isolate the group (or remove the group from possible fault) by a single test at the input or output test point for the group.

For example, there are three circuit groups (frequency converter, IF amplifier, and audio amplifier) and two individual circuits (RF amplifier and detector) in the receiver of Fig. 1-4. Each of the three groups and the two individual circuits have input/output test points.

The input circuit for the complete functional unit is the RF amplifier Q_1, and the base of this transistor is the input signal-injection point. The output from the complete unit is at the secondary of the transformer. The input is where the functional unit (receiver) receives a signal from another unit (antenna switch), and the output supplies a signal to another unit or device (the loudspeaker).

This same idea applies to a circuit group. That is, the input signal to the group is injected at one point, and then the output signal is obtained at a point *several stages* farther along the signal path.

Note that the IF amplifier circuit group is an integrated circuit, consisting of three amplifier stages. Although each stage has an input and an output, none of the stages has separate input/output test points, as do the various stages in the audio amplifier circuit group (which is made up of replaceable parts). Even if it were possible, there would be no point in testing each stage of the IC IF amplifier. The entire IC is replaced as a package.

In most cases the input signal is injected at the base, and the output signal is traced at the collector (or possibly the emitter). That is, the input is at the first base in the signal path, while the output is at the last collector in the same path. There are certain exceptions to this. For example, in the frequency converter circuit group, the RF oscillator Q_3 has no input, but it does provide an output to the mixer stage Q_2. Also, the detector stage CR_1 is a diode, with the input at the anode and the output at the cathode.

To determine the signal-injection and output points of a circuit group, you must find the *first circuit* of the group in the signal path and the *final circuit* of the group in the path. For example, the input for the audio amplifier circuit group is at the base of Q_4, while the output is at the secondary of transformer T_{10}. Note that test point D is at the input of the audio amplifier circuit group as well as at the output of the detector circuit.

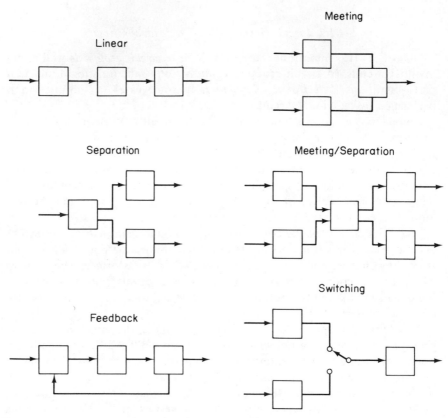

Figure 1–5. Types of signal paths.

1–4.2. Signal Paths

There are six basic types of signal paths, no matter what circuit group or circuit arrangement is used. These types are shown in Fig. 1-5 and are summarized as follows.

A *linear path* is a series of circuits arranged so that the output of one circuit feeds the input of the following circuit. Thus, the signal proceeds straight through the circuit group without any return or branch paths.

A *meeting path* is one in which two or more signal paths enter a circuit.

A *separating path* is one in which two or more signal paths leave a circuit.

A *meeting/separating* is one in which a single stage has multi-inputs and multi-outputs.

A *feedback path* is a signal path from one circuit to a point or circuit *preceding* it in the signal flow sequence.

A *switching path* contains a selector switch (or similar device such as a relay) that provides a different signal path for each switch position.

The main signal path through the IF amplifiers is an example of a linear path. The automatic volume control (AVC) line from the detector to the IF amplifiers is an example of the feedback path. The paths to the mixer Q_2 and the output transformer T_{10} are both meeting paths. The paths from the transmitter and receiver to the antenna assembly (through the antenna switch, Fig. 1-3) are switching paths.

1-4.3. Signal Tracing Versus Signal Substitution

Both signal-tracing and signal-substitution (or signal-injection) techniques are used in solid-state and digital troubleshooting.

Signal tracing is accomplished by examining the signal at a test point with an oscilloscope, multimeter, loudspeaker, and the like. In signal tracing, the input probe of the indicating device used to trace the signal is moved from point to point while the signal is applied at a fixed point, either from an internal or an external source.

Signal substitution or injection is accomplished by injecting an artificial signal (from a signal generator, sweep generator, etc.) into a circuit or a complete functional unit to check its performance. In signal injection, the injected signal is moved from point to point while the input probe to the indicating device remains fixed at one point.

Troubleshooting often involves *both signal tracing and signal substitution*. For example, when troubleshooting a solid-state audio system, an audio oscillator can be used to inject a signal at the input, while an oscilloscope can be used to observe the waveform at each stage or circuit. Or, a repetitive pulse can be introduced into a digital system by means of a pulse generator. The resultant waveforms at various gates, flip-flops, delays, and so on, are then observed on an oscilloscope.

1-4.4. Half Split Technique

The half-split technique is based on the idea of simultaneous elimination of the maximum number of circuit groups or circuits with any test. This will save both time and effort. The half-split technique is used primarily when isolating trouble in a linear signal path. That is, after brackets have been placed at good input and bad output points, unless the symptoms *point definitely* to one circuit group or circuit which might be the trouble source, the most logical place to make the first test is at a *convenient* test point *halfway between* the brackets.

1-4.4.1 Example of the Half-Split Technique

The block diagram shown in Fig. 1-6 is a simplified version of the receiver servicing block diagram of Fig. 1-4. Figure 1-6 illustrates the

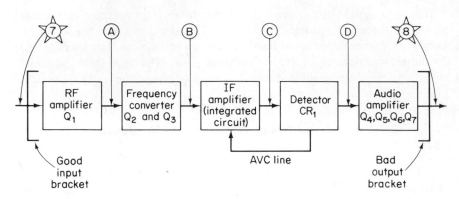

Figure 1–6. Simplified servicing block diagram of receiver unit illustrating the linear signal path of received signal through circuit groups.

linear signal path of the received signal through the receiver unit by showing the circuit groups consolidated into single blocks. The brackets placed at test point 7 (good input) and test point 8 (bad output) show the trouble being localized to the receiver unit.

The next phase of troubleshooting is to isolate the trouble to one of the circuit groups (frequency converter, IF amplifier, or audio amplifier) or one of the individual circuits (RF amplifier or detector) in the linear signal path.

Assuming that test points A, B, C, and D are equally accessible and that there are no special symptoms that would point to a particular circuit or group, test point C is the most logical point for the first test if *signal tracing* is used, with test point B the next logical choice. If *signal substitution* is used, test point D is the most logical choice. The troubleshooting sequence with signal tracing will be discussed first.

With signal tracing, an RF signal (at the receiver frequency) modulated by an AF tone is introduced at test point 7. An oscilloscope is then connected to monitor the waveform at the various test points.

If test point A is chosen first and the oscilloscope display is normal, the trouble is located somewhere between A and the receiver output (test point 8). This means that the trouble could be in the frequency converter, IF amplifier, detector, or audio amplifier. On the other hand, if an abnormal signal is observed at test point A, the trouble is then immediately isolated to the RF amplifier. All other circuits are eliminated. This would be a lucky break. However, to be performed efficiently and rapidly, the troubleshooting procedures should be based on a systematic, logical process and not on chance or luck.

The same condition is true if test point D is chosen first. A normal oscilloscope display will clear all circuits but the audio section. However, an abnormal display will still leave the possibility of trouble in many circuits.

If test point B is chosen first and the oscilloscope display is normal, this will clear three circuits (RF amplifier, mixer, RF oscillator) but will leave six circuits possibly defective (the IC IF amplifier, detector, audio voltage amplifier, phase splitter, and two audio power amplifiers). The opposite results will be obtained if the oscilloscope display is abnormal.

If test point C is chosen first and the oscilloscope display is normal, this will clear four circuits (RF amplifier, mixer RF oscillator, and the IC IF amplifier) but will leave five circuits possibly defective. This process divides the circuits into two groups (known good and possibly bad).

Now assume that there is an abnormal indication at test point C. The bad output bracket can be moved to test point C, with the good input bracket remaining at test point 7. The next logical signal-tracing test point is B, since this is now near the halfway point between the two brackets. If test point A is chosen before test point B, it will confirm or deny the possibility of trouble in the RF amplifier *only*.

If the oscilloscope display is normal at test point B but abnormal at C, the trouble is isolated to the IF amplifier. Since the IF amplifier is an IC, it must be replaced as a unit and no further trouble isolation is required. (There is an exception, however, that should be checked before going to the trouble of replacing the IC. It is possible that the AVC feedback from the detector to the IF amplifiers is absent or abnormal, causing the IF amplifier to produce an abnormal display at C.)

If the oscilloscope display is abnormal at test point B, the trouble is isolated to the RF amplifier or the frequency converter. Additional observation at test point A will further isolate the trouble to either circuit.

If there is an abnormal indication at test point A, then the bad output bracket can be moved to A, and trouble is isolated to the RF amplifier. If there is a normal indication at A, the good input bracket can be moved to A, and trouble is isolated to the frequency converter.

With *signal injection,* signals of the right sort are injected at tests points A, B, C, and D. Receiver response is noted on the loudspeaker. In this case, the signal for test point A is at the RF frequency of the receiver, modulated by an AF tone. The signals for B and C are at the IF frequency, also modulated by an AF tone. The signal for D is an AF tone.

Using signal injection with the half-split technique, the first signal is again injected at test point C. Now, however, a normal response from the loudspeaker (with a signal at C) will clear the final five circuits (detector and four audio circuits). Under these circumstances the next logical points for signal injection are B and A, in that order.

An absent or abnormal response (with a signal injected at C) will isolate the trouble to the detector and audio circuits. Under these circumstances the next logical test point is D.

A normal response from the loudspeaker with a signal at D but not at C, will isolate the trouble to the detector. An absent or abnormal response with a signal at D isolates the trouble to the audio circuits.

Keep in mind that the preceding examples using the half-split technique, signal tracing, signal substitution, and bracketing to isolate the trouble to a circuit group by no means cover all the possibilities that may occur. The examples were used only to illustrate the basic concepts involved when following the systematic, logical, troubleshooting procedure.

1–4.5. Isolating Trouble to a Circuit Within a Circuit Group

Once trouble is isolated to a faulty circuit group in the receiver unit, the next step is to isolate the trouble to the faulty circuit within the group. Bracketing, half-splitting, signal tracing, signal substitution, and knowledge of signal path in the circuit group are all important to this step and are used in essentially the same way as for isolating trouble to a circuit group.

Isolating the trouble to a circuit in a linear path is identical to that described in the preceding sections. However, isolating trouble to a circuit in a separating or meeting signal path requires a variation in the procedure. Although the separating or meeting signal paths are not recognized as easily as linear paths, you should have little difficulty in recognizing them if you follow the definitions given in Sec. 1–4.2.

Sometimes, various types of signal paths are combined. For example, as shown in Fig. 1–7, the audio circuit group of the receiver contains a linear path (from audio voltage amplifier to phase splitter), a separating path

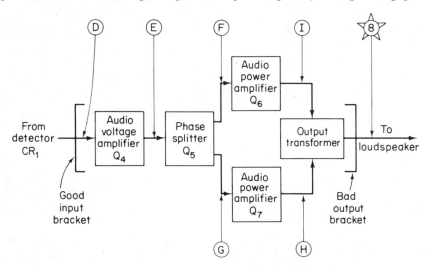

Figure 1–7. Receiver audio section showing separating and meeting paths.

(phase splitter to power amplifiers), and a meeting path (power amplifiers to output transformer).

1–4.5.1. Example of Isolating Trouble to a Circuit (Separating and Meeting Paths)

In Fig. 1–7 the audio amplifier circuit group is illustrated in a manner slightly different from the servicing block diagram of Fig. 1–4. The (good input) bracket at test point D and the (bad output) bracket at test point 8 indicate that the trouble is isolated to the circuit group. There is a normal signal (injected from an audio signal generator) at the input base of the audio voltage amplifier Q_4, and no signal is observed (on an oscilloscope) at the output transformer.

Before you can isolate the trouble in separating or meeting signal paths, you must first isolate these paths from the rest of the circuit group. That is, the phase splitter, audio power amplifiers, and output transformer must first be isolated from the audio voltage amplifier. This is done by making the first test at point E.

Assume that the oscilloscope is now moved to test point E and a normal signal is observed there. This means that the voltage amplifier is functioning normally. The good input bracket is moved to test point E, and trouble is isolated to the phase splitter, power amplifiers, or output transformer. (If the signal had been abnormal at E, the trouble is isolated immediately to the audio voltage amplifier.) With a normal signal at E, the trouble is isolated to the separating or meeting signal paths in the circuit group.

The next step is to eliminate one of the paths and thus isolate the trouble to the other path. For troubleshooting purposes, the remaining circuits can be considered as *two linear paths:* (1) phase splitter Q_5, power amplifier Q_6, and transformer; and (2) phase splitter Q_5, power amplifier Q_7, and transformer. This involves checking signals at test points F, G, H, and I.

The next check is made at test point H or I. If the next check is made at F or G and a normal signal is observed, nothing is proved.

If the signal is normal at I, the top signal path is eliminated and the next check should be made at H. If the signal is abnormal at H, the bad output bracket is moved to H. Then test point G is the next logical point for a check. If the indication at G is normal, the good input bracket is moved to G and trouble is isolated to power amplifier Q_7. If the indication at G is abnormal, the bad output bracket is moved to G and trouble is isolated to phase splitter Q_5.

If the signal is abnormal at I, trouble is isolated to the top signal path and test point F is the logical point for a check. If the indication at F is normal, the good input bracket is moved to F and trouble is isolated to power am-

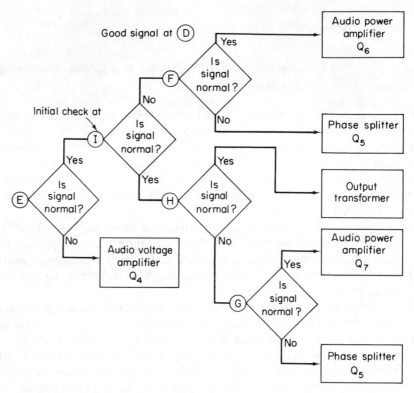

Figure 1–8. Thought process for isolating trouble in audio circuit group (initial test made at I, after test at E).

plifier Q_6. If the indication at F is abnormal, the bad output bracket is moved to F and the trouble is isolated to phase splitter Q_5.

The thought process involved for isolating trouble in the audio circuit group is shown in the flow diagrams of Figs. 1–8 and 1–9. The diagrams are similar to those used in the troubleshooting section of some technical manuals. Figure 1–8 is based on the assumption that the initial test is made at test point I, after the test at E. Figure 1–9 is based on an initial test at H, after the test at E.

1–4.5.2. *Example of Isolating Trouble to a Circuit (Feedback Path)*

As discussed, a feedback path is a signal path from one circuit to a point or circuit preceding it in the signal-flow sequence. In the receiver under consideration, a feedback path is represented by the AVC line from the detector CR_1 to the bases of the IF amplifiers in the IC module, as illustrated in Fig. 1–10.

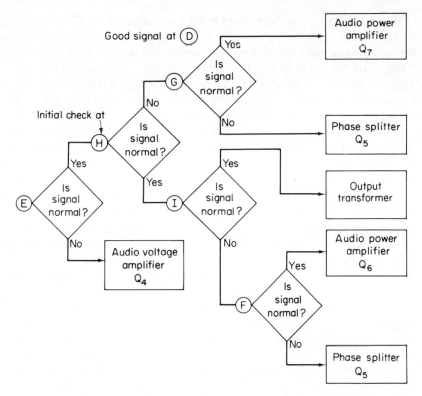

Figure 1–9. Thought process for isolating trouble in audio circuit group (initial test made at H, after test at E).

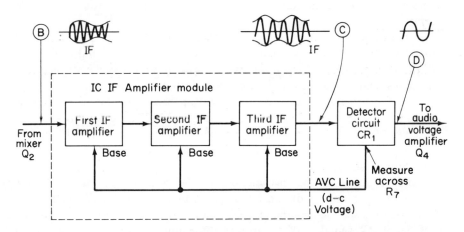

Figure 1–10. Example of feedback path (AVC line).

When a strong signal is received from the mixer, the AVC voltage output of the detector prevents the audio (main signal flow) from becoming too strong. A weak input signal, however, does not result in a high AVC voltage. Thus, there is enough gain in the circuit to amplify the weak signal.

If the trouble is isolated to the circuits of Fig. 1–10, you should first examine the output of the detector to find out if the fault is most likely in the main signal path or in the feedback path.

If the detector output is weak or if there is no output whatsoever when the input to the IF module is normal (at test point B), the half-split method can be applied to the main signal path. That is, the signal should be measured at test point C.

On the other hand, an output signal which is considerably stronger than normal for given input is an indication that the fault is in the feedback path. In this case the AVC voltage output of the detector should be checked first.

Note that the AVC line is not assigned a test point number or letter. However, there is a note indicating that the AVC voltage can be measured across R_7 in the detector circuit. A similar note or a test point assignment will be found in technical manuals or data sheets. In well-prepared service literature, the physical location of test points and important circuit parts is given by means of photographs or drawings. Unfortunately, in some service literature, you must find the electrical location on the schematic diagram and then hunt for the physical location, possibly with the aid of an illustrated parts list.

A low AVC voltage indicates that the trouble is in the detector circuit. In solid-state equipment the AVC feedback voltage can be positive or negative, depending on whether the IF amplifier transistors are PNP or NPN. Either way, the AVC output must provide an opposing voltage having a value high enough to decrease the gain of the IF amplifiers when a strong signal is received.

If the opposing voltage is low or if there is no AVC voltage whatsoever, there will be no feedback signal. Consequently, a strong signal at the input of the IF amplifiers will be amplified at a high level, causing a stronger than normal signal at the audio output of the detector.

The thought process involved for isolating trouble in the AVC feedback circuit is shown in the flow diagram of Fig. 1-11. The diagram is similar to those in the troubleshooting section of some technical manuals. Figure 1-11 is based on the assumption that the initial test is made at test point D (the main audio signal path).

Another method (besides measuring the voltage) of troubleshooting circuits that contain feedback paths is to *disable the feedback loop*. This may be done by disconnecting the feedback path (opening the loop) or by shorting the feedback signal path to ground. Opening the loop is sometimes inconvenient, and shorting the AVC signal to ground could cause damage to

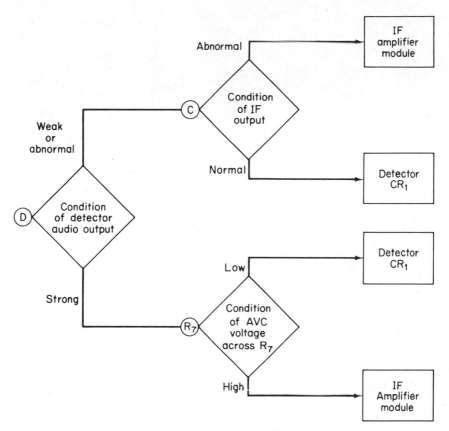

Figure 1–11. Thought process for isolating trouble in AVC feedback circuit (initial test made at D).

the circuit. *Never short any circuit to ground unless you are certain that no damage will result.*

One method for disabling a feedback line without damage is to lift the line at the source and connect a resistance between the source and ground. In this case, you disconnect the line from the detector, measure the resistance of the line (to the IF module), and then connect an equivalent resistance between the source (detector circuit) and ground. In this way you can disable the AVC line and measure the available AVC voltage across the equivalent resistance.

Note that some receivers (particularly communications-type receivers) are provided with an AVC *on/off* switch. If trouble is located in the circuit containing the AVC loop, the switch can be used to find out if the trouble still exists without AVC.

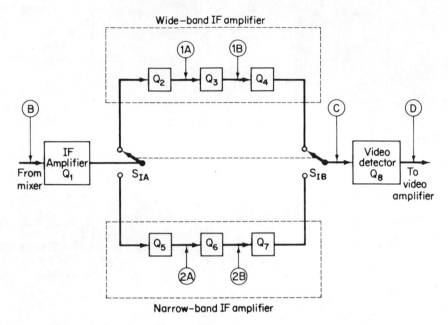

Figure 1–12. Example of switching circuit signal paths (IF circuits of radar receiver).

1–4.5.3. Example of Isolating Trouble to a Circuit (Switching Circuit)

To isolate the faulty circuit in a switching path, you initially test the final output for the circuit *following the switch, as the switch positions are changed.* When the switch is a multiple-contact type, each contact may be connected to a different circuit. In this case it may be necessary to place the switch in each position and check the final output of the circuit associated with that position. If the symptoms and/or tests point to one specific circuit, it may not be necessary to check every switch position. Once this test has been performed and the trouble isolated to one or more of the branches, the suspected branches are then checked to locate the one that is faulty. The next step is to apply the techniques of half-splitting and isolating —as for linear, separating, meeting, or feedback paths— to isolate the faulty circuit.

Since a switch connects different combinations of circuit groups, the switch can be used to help isolate the trouble to one or more of the circuits.

For example, consider the switching path of Fig. 1–12, which shows the IF circuits of a radar receiver. As shown, there are two separate IF signal paths, wide-band and narrow-band, which can be selected by switch S_1. The two sections of S_1 are ganged together.

By observing the output of video detector Q_8 (test point D) with a proper IF signal injected at the input of the first IF amplifier Q_1 (test point B), switch S_1 can be used to help isolate the trouble.

If the output of Q_8 is incorrect *only during wide-band operation,* the trouble is in the wide-band circuit group Q_2, Q_3, and Q_4. If the output of Q_8 is incorrect *only during narrow-band operation,* the trouble is in the narrow-band circuit group Q_5, Q_6, and Q_7. If the abnormal symptoms are present for *both positions of the switch,* the trouble is in the first IF amplifier Q_1, or in the video detector Q_8, or in the switch S_1 itself.

The rest of the troubleshooting procedure consists of narrowing down the trouble to a single circuit within the faulty circuit group.

For example, assume that a check at test point D shows an abnormal signal for both positions of S_1 when a normal signal is injected at test point B. The next logical place for a signal-tracing test is at test point C.

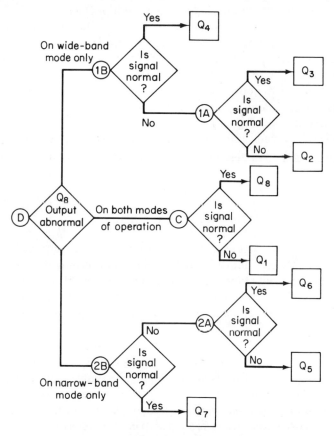

Figure 1–13. Thought process for isolating trouble in switching path (good signal at B, initial test made at D).

Since both the narrow-band and the wide-band modes of operation are abnormal at the video detector output, the trouble must be in a circuit common to both modes. In this case the trouble is in either the video detector Q_8 or the first IF amplifier Q_1. Checking for a signal at test point C will isolate the trouble to one of these two circuits. All of the circuits in the narrow-band and wide-band circuit groups (Q_2 through Q_7) are eliminated because a signal in any one of these circuits will produce an abnormal signal for only *their respective mode of operation*.

With a normal signal at C, the trouble is isolated to the video detector Q_8. An abnormal signal at C (in both positions of the switch) establishes the trouble at the first IF amplifier Q_1.

The thought process involved for isolating trouble in the switching circuit is shown in the flow diagram of Fig. 1–13. This diagram is similar those shown in some technical manuals. Figure 1–13 is based on the assumption that the initial test is made at test point D (with a good input signal at B).

For conditions where the trouble is isolated to the narrow-band or the wide-band circuit groups, the flow diagram illustrates the respective first tests being made at the output of the second IF amplifier, Q_3 or Q_6 depending on the mode of operation. The flow diagram will be slightly different if the first test is made at the output of the first amplifier in each case. However, the thought process and logic are identical.

The test sequence of Fig. 1–13 does not include testing for a faulty switch S_1. Should trouble be isolated to a circuit which is connected to one of the switch contacts, a check should be made between the circuit and the switch before the circuit is bracketed.

1–5. LOCATING A SPECIFIC TROUBLE

The ability to recognize symptoms and to verify them with test equipment will help you make logical decisions concerning the selection and localization of the faulty functional unit and isolate trouble to a faulty circuit. The final step of troubleshooting—to locate the specific trouble—requires testing of the various branches of the faulty circuit to find the defective component.

Because so much solid-state equipment is of IC and sealed-module design, technicians often assume that it is unnecessary to locate specific troubles to individual components. That is, they assume all troubles can be repaired by replacement of sealed modules. This is not true. While the use of replaceable modules often *minimizes the number of steps required* in troubleshooting, it is still necessary to check circuit branches to components outside the module. Front panel operating controls are a good example of this, since such controls are not located in the sealed units. Instead, they are connected to the terminals of an IC, circuit board, or plug-in module.

The proper perfomance of the *locate* step will enable you to find the cause of trouble, repair it, and return the equipment to normal operation. A follow-up to this step is to record the trouble so that, from the history of the equipment, future troubles may be easier to locate. Also, such a history may point out consistent failures which could be caused by a design error.

1–5.1. Inspection Using the Senses

After the trouble is isolated to a circuit, the first step in locating the trouble is to perform a preliminary inspection using the senses, particularly those of sight, smell, hearing, and touch. For example, burned or charred resistors can often be spotted by visual observation or by smell. The same holds true for oil-or wax-filled components, such as capacitors, inductors, and transformers. When overheated, the oil or wax in these components expands and usually leaks out or perhaps causes the case to buckle or even explode. Overheated components, such as hot transistor cases, can be located quickly by touch. The sense of hearing can be used to listen for high-voltage arcing between wires or between wires and the chassis, for "cooking" of overloaded or overheated transformers, or for hum or lack of hum—whichever the case may be. Although the senses of sight, smell, hearing, and touch are used at this time, this procedure is referred to more frequently as a *visual inspection*.

Although it is possible to find a defective component by visual inspection, the component should not be replaced until the trouble has been investigated further. Often, the real cause of trouble will destroy another component. For example, a short can cause a resistor to burn out. If the resistor is replaced without first removing the short, the new resistor will also burn out. *So always check for other possible troubles after locating a defective component.*

1–5.2. Testing to Locate a Faulty Component

Unlike vacuum tubes, transistors and solid-state diodes are not easily replaced. Therefore, the old electronics troubleshooting procedure of replacing tubes at the first sign of trouble has not carried over into solid-state equipment. Instead, solid-state circuits are analyzed by testing to locate faulty components.

For service and maintenance purposes, the transistor and semiconductor diode may be considered the *common denominator* in an electronic circuit. Because of their key position in the circuit, these devices are a convenient means of evaluating the operation of the circuit (through waveform, voltage, and resistance measurements) and locating trouble in a relatively short time.

Usually, the first step in solid-state circuit testing is to analyze the *output waveform*, if there is one. That is, while isolating the trouble to the circuit, assume that an abnormal or distorted output waveform was noted. This waveform is now analyzed in detail to check the amplitude, duration, phase, and/or shape to aid in making a valid deduction of the probable branch (collector, emittter, or base) of the circuit in which the trouble may be.

It is possible to test transistors and diodes in circuit, using in-circuit testers. Such testers are discussed further in Chapter 2. These testers are usually quite good for transistors used at lower frequencies, particularly in the audio range. However, most in-circuit transistors will not show the *high-frequency* and/or *switching* characteristics of transistors. (The same is true of out-of-circuit transistor and diode testers.) For example, it is quite possible for a transistor to perform well as an audio amplifier but be hopelessly inadequate as a high-speed switching device required in digital equipment.

After a waveform analysis and/or in-circuit test of transistors and diodes, the next logical step is voltage measurement at the transistor and diode terminals or leads. When a properly prepared technical manual or data sheet is available (with waveform, voltage, and resistance charts), the voltage measurements can be compared with the *normal* voltages listed in available voltage charts. This check will often help isolate the trouble to a single branch of the circuit.

It is often necessary to troubleshoot solid-state circuits without benefit of normal voltage and resistance information. This can be done using the schematic diagram and logic. For example, with an NPN transistor, the base must be positive in relation to the emitter if there is to be emitter-collector current flow. That is, the emitter-base junction will be forward-biased when the base is more positive (or less negative) than the emitter. The problem of troubleshooting solid-state circuits on the basis of relative voltages is discussed further in Chapter 2.

After voltage measurements are made, it is often helpful to make resistance measurements at the same points, particularly where an *abnormal voltage measurement* is found. Suspected parts often can be checked by a resistance measurement, or a continuity check can be made to find the point-to-point resistance of the suspected branch. As discussed, considerable care must be used when making resistance measurements in solid-state circuits. The junctions of a transistor act like diodes. When properly biased (say, by the ohmmeter battery) the diodes will conduct and produce false resistance readings. This condition is discussed in Chapter 2.

In rare cases, the current in a particular circuit branch must be measured directly with an ammeter. However, it is usually simpler and more practical to measure the voltage and resistance of a circuit, and then calculate the current.

1–5.3. Voltage Measurements

When testing to locate trouble, the voltage measurements are made with the circuit in operation but with no signal applied. If you are using voltage charts in a manual or data sheet, follow all of the notes and precautions on the charts. Usually, the charts will specify the position of operating controls, typical input voltage, and so on.

Because of the safety practice of setting a voltmeter to its highest scale before making measurements, the element having the highest voltage— usually the collector—should be checked first. Then the elements having lesser voltage should be checked in descending order, that is, the emitter and then the base. Of course, in some circuits the emitter is at ground.

If you have had any *practical experience* in troubleshooting, you know that voltage (as well as resistance and waveform) readings are seldom identical to those listed in the technical manuals and data sheets. (The same is true of troubles listed in the technical manual trouble charts. You will rarely find trouble in actual maintenance that appears in the manual.)

This brings up an important question concerning voltage checks: "How close is good enough?" In answering this question, there are several factors to consider.

The tolerances of the resistors, which greatly affect the voltage readings in a circuit, may be 20, 10, or 5 per cent. Resistors with 1 per cent tolerance are used in some critical circuits. The tolerances marked or color-coded on the parts are therefore one important factor. Transistors and diodes have a fairly wide range of characteristics and thus will cause variations in voltage readings.

The accuracy of test instruments must also be considered. Most voltmeters have accuracies of 5 to 10 per cent, while precision meters are more accurate.

For proper operation, critical circuits may require voltage readings to within 10 per cent of the values specified in the manual. However, most circuits will operate satisfactorily if the voltages are to within 20 or 30 per cent.

Generally, the most important factors to consider are the symptoms and the output signal. If no output signal is produced, you should expect a fairly large variation of voltages in the trouble area. Trouble which results in a circuit performance, just out of tolerance, may cause only a slight change in circuit voltages.

1–5.3.1. Duplicating Voltage Measurements

If you are responsible for the maintenance or service of one piece of electronic equipment or one type of equipment, it is *strongly recommended* that you duplicate all of the voltage measurements found in the technical manual with test equipment of your own. This should be done when the equipment is operating properly. (The same applies to waveform and resistance measurements.) Then when you make voltage measurements, you can spot even slight variations in voltage. Always make the initial measurements with test equipment that will be used during troubleshooting procedures. If more than one set of test equipment is used, make the initial measurements with all available test equipment and *record the variations.*

1–5.4. Resistance Measurements

Unlike voltage measurements, which are made with the equipment turned on, resistance measurements must be made with no power applied. However, in some cases various operating controls must be in certain positions to produce resistance readings similar to those found in the manual resistance charts. This is particularly true of controls that have variable resistances. Always observe any notes or precautions found on the manual resistance charts. In any circuit, another safety feature to protect the ohmmeter is to ensure that all filter capacitors are discharged.

After these items are checked, the resistance measurements can be made from the transistor elements to the chassis (ground) or between any two points that are connected by wiring or parts.

Because of the shunting effect of other components connected in parallel, the resistance of an individual component or circuit may be difficult to check. In such cases, it is necessary to disconnect one terminal of the component being tested from the rest of the circuit. This will leave the component open at one end, and the value of resistance measured will be for that component only.

Keep in mind that while making a resistance check, a zero reading indicates a short circuit, and an infinite reading indicates an open circuit. Also remember the effect of the transistor junctions (acting as a forward-biased diode when biased properly). Refer to Chapter 2.

1–5.5. Using Schematic Diagrams

Regardless of the type of trouble symptom, the actual fault can be traced eventually to one or more of the circuit components—transistors, diodes, resistors, capacitors, coils, transformers, and so on. The voltage and/or resistance checks will indicate which branch within a circuit is at fault. Then you must locate the particular component that is causing the trouble in the branch.

This requires that you be able to read a schematic wiring diagram. These diagrams show what is inside the blocks on a servicing block diagram and provide the final picture of the electronic equipment. Often you must service

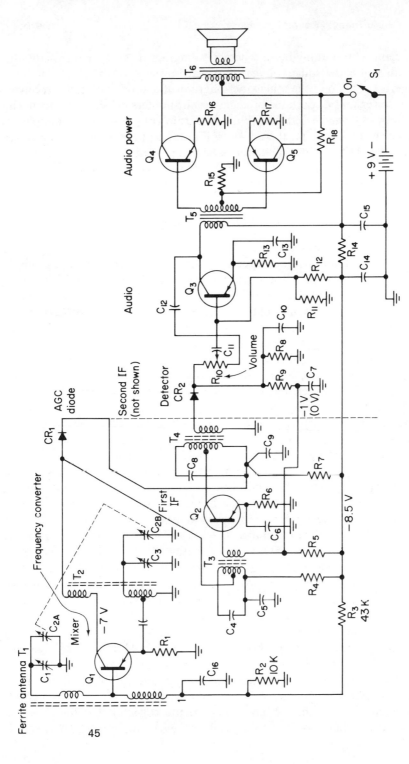

Figure 1-14. Typical broadcast band solid-state receiver (transistor portable) schematic diagram.

45

equipment with nothing but a schematic diagram. If you are fortunate, the diagram will show some voltages and waveforms.

Figure 1–14 shows the schematic diagram of a solid-state radio receiver. This receiver unit (a portable broadcast radio) differs considerably from the one illustrated in the servicing block diagram (Fig. 1–4). For example, in the receiver of Fig. 1–14, there is no RF amplifier; the frequency converter (mixer and RF oscillator) function is accomplished by one transistor, Q_1; there are only two IF amplifiers (one not shown); and there is no phase splitter in the audio section.

1–5.5.1. Examples of using Schematic Diagrams

The following examples show how the schematic diagram is used to locate a fault within a circuit. While these examples involve a simple transistor portable radio, the same *basic troubleshooting principles* apply to more complex solid-state equipment.

Example No. 1. Assume that the receiver of Fig. 1–14 is being serviced and trouble has been isolated to the frequency converter. A signal of appropriate frequency (IF) modulated by an audio tone is injected at the IF input (primary of transformer T_3), and a good response is heard on the loudspeaker. However, no response is obtained when a signal of proper frequency (RF) is injected at the primary of T_1 with the receiver tuned to that frequency.

The next logical step is to measure the voltages at the collector, emitter, and base of Q_1 (in that order). If any of the Q_1 elements show an abnormal voltage, the resistance of the element should be checked first. Note that the collector voltage is specified as -7 V. The base and emitter voltages are not given, nor are any of the resistance values. This lack of information is typical of solid-state equipment used in the home-entertainment field. Therefore, you must be able to interpret schematic diagrams to *estimate approximate voltages*. For example, the voltage at the junction of R_3 and R_4 is given as -8.5 V. This is logical since the source voltage is -9 V. The value of R_2 is approximately 25 per cent of R_3. Therefore, the drop across R_2 is about 25 per cent of -8.5 V or approximately -2 V. If Q_1 is silicon, the emitter will be about 0.5 V different from the base (emitter more positive or less negative, in this case). Q_1 is germanium, the base-emitter differential will be about 0.2 V.

Keep in mind that this method of interpreting the schematic will give you *approximate voltages only.* For practical troubleshooting, the voltage differentials between circuit elements and transistor electrodes are the most important factor. Solid-state troubleshooting based on voltage differentials is discussed fully in Chapter 2.

Now assume that there is no voltage at the collector, but the base and emitter show what appear to be normal or logical voltages. The next step is to remove power, discharge C_{14} and C_{15} (if necessary), and measure the collector resistance (to ground). Since there are no resistance values given, you must use the schematic to estimate the approximate values.

Of course, if you find a zero resistance at the collector, this indicates a short. For example, capacitor C_5 could be shorted. On the other hand, an infinite resistance indicates an open circuit. For example, the coil windings of T_2 and T_3 could be open, or R_4 could be burned out and open. It is usually easy to locate the fault when you find such extreme resistance readings.

However, a resistance reading between these two extremes does not provide a really sound basis for locating trouble. To make the problem worse, the effect of solid-state devices in the circuit can further confuse the situation.

For example, assume that the fault is an open T_3 winding. This will result in a no-voltage reading at the collector of Q_1. However, it is still possible to measure a resistance to ground from the Q_1 collector. Assume that the ohmmeter leads are connected so that the *positive* terminal of the ohmmeter battery is connected to the Q_1 collector. This will forward-bias the AGC diode CR_1 and cause the ohmmeter to measure the resistance across R_7 (the collector supply of Q_2). Also, if the collector of Q_1 is made positive in relation to the base, the Q_1 base-collector junction will be forward-based, resulting in possible current flow.

This problem can be eliminated by reversing the ohmmeter leads and measuring the resistance both ways. If there is a difference in the resistance values with the leads reversed, check the schematic for possible forward-biased diode or transistor junction.

Example No. 2. Assume that the receiver of Fig. 1–14 is being serviced and trouble is isolated to the detector. An audio tone injected at the base of Q_3 produces a good response on the loudspeaker. However, no response is obtained when an IF signal modulated by an audio tone is injected at transformer T_4. Or, with signal tracing instead of signal injection, an oscilloscope connected across the secondary of T_4 shows that an AF modulated signal is available to the detector, but no AF signal appears at the base of Q_3.

The detector CR_2 has two outputs: an AF signal and an AGC voltage. Under no-signal conditions, electrons flow through R_9 and R_8 to ground, producing approximately -1 V bias at the junction of R_5 and R_9. With a normal signal, electrons flow through R_8 and CR_2. This flow opposes the normal bias and produces approximately 0 V at the junction of R_5 and R_9.

The AGC circuit function can be checked by measuring the voltage on the AGC line, with or without a signal applied to T_4. If there is an approximately 1 V change with signal, the AGC function *and detector* CR_2 can be considered normal.

With CR_2 established as normal but with no signal at the base of Q_3, volume control R_{10} or coupling capacitor C_{11} are logical suspects, probably open or with broken leads. Capacitor C_{12} can also be a possible suspect, but it is not as likely as C_{11} or R_{10}. If C_{12} is shorted or leaking, this will show up as an abnormal voltage (a large negative voltage on the AGC line). If C_{12} is open, the frequency response characteristics of the audio circuit will be poor, but there will still be a signal at the base of Q_3.

The condition of C_{11} can be checked either of two ways (without removal from the circuit). First, a signal (AF) can be injected on both sides of C_{11}. If the signal is heard on the loudspeaker when injected at the base of Q_3 but is not heard when injected at the junction of R_{10} and C_{11}, capacitor C_{11} is probably defective. Second, a signal can be injected ahead of the detector and traced on both sides of C_{11} (with an oscilloscope or signal tracer). The signal should appear substantially the same on both sides of C_{11}. If not, C_{11} is probably defective.

The condition of R_{10} can be checked by signal tracing or signal injection as well. However, as a final test, the power should be removed and the resistance of R_{10} measured. To measure potentiometer resistance, connect the ohmmeter from the wiper arm to one end of the winding; then vary the control from one extreme to another. Repeat the test with the ohmmeter connected from the wiper arm to the opposite end of the winding. The resistance indication should vary smoothly with no jumps or flicking of the ohmmeter needle. Such jumps can mean bad spots (open or poor contact) on the potentiometer winding.

1–5.6. Internal Adjustments During Trouble Localization

Keep in mind that adjustment of controls (both internal adjustment controls and operational controls) can affect circuit conditions. This may lead to false conclusions during troubleshooting.

For example, the amplitude of the signal at the base of Q_3 (Fig. 1–14) is set directly by the volume control (R_{10}), which is an operational control. The amplitude of both the signal and AGC voltage from the detector can be affected by adjustment (or alignment) of the IF transformers (as well as the oscillator and RF transformers).

If the signal at Q_3 is very low, it *could be* that the volume control is set to minimum. Of course, since the volume control is an operational control, a run-through of the operating sequence at the beginning of troubleshooting will pinpoint such an obvious condition. However, a low output from the detector *could be* caused by poor alignment of the IF transformers. Since the IF transformers require internal adjustments, poor alignment will not be found by operating procedures. It will require adjustment of the IF transformers.

This condition, or a similar one, leads some technicians to one of two *unwise* courses of action:

First, the technician may launch into a complete alignment procedure (or whatever internal adjustments are available) once he has isolated the trouble to a circuit and is trying to locate the specific defect. No internal control no matter how inaccessible is left untouched! He reasons that it is easier to make adjustments than to replace parts. While such a procedure will eliminate improper adjustment as a possible fault, it can also create more trouble than is repaired. Indiscriminate internal adjustment is the technician's version of operator trouble.

Second, a technician may replace part after part, when a simple screwdriver adjustment will repair the problem. This attitude can be caused by the technician's inability to perform the adjustment procedures or his lack of knowledge concerning the control's function in the circuit. Either way, a study of the service literature should resolve the situation.

To take the middle ground, *do not make any internal adjustments during the troubleshooting procedure until trouble has been isolated to a circuit, and then only when the trouble symptom or test results indicate possible maladjustment.*

For example, assume that an oscillator is provided with an internal adjustment control that sets the frequency of oscillation. If waveform measurement at the circuit shows that the oscillator is off frequency, it is logical to adjust the frequency control. However, if waveform measurements show only a very low output (but on frequency), adjustment of the frequency control during troubleshooting could cause further problems.

An exception to this rule is when the service literature recommends alignment or adjustment as part of the troubleshooting procedure. Generally, alignment or adjustment is checked after test and repair have been performed. This assures the repair (parts-replacement) procedure has not upset circuit adjustment.

1–5.7. Trouble Resulting from More than One Fault

A review of all the symptoms and test information obtained thus far will help you verify the component located as the sole trouble or isolate other faulty components, whether the malfunction of these components is due to the isolated component or to some entirely unrelated cause.

If the isolated malfunctioning component can produce *all the normal and abnormal symptoms and indications* that you have accumulated, you can logically assume that it is the sole cause of the trouble. If not, then you must use your knowledge of electronics and of the equipment to find what other component or components could have become defective and produce all of the symptoms and test data.

When one component fails, it often results in abnormal voltages or currents which could damage other components. Trouble is often isolated to a faulty component which is a result of the original trouble rather than its source.

For example, assume that the troubleshooting procedure thus far has isolated a transistor as the cause of trouble—the transistor is burned out. What would cause this? Excessive current can destroy the transistor by causing internal shorts or by altering the characteristics of the semiconductor material, which is very temperature sensitive. Thus, the problem becomes a matter of finding how excessive current can be produced.

Excessive current in a transistor circuit can be caused by an extremely large input signal, which will overdrive the transistor. Such an occurrence will indicate a fault somewhere in the circuitry ahead of the input connection.

Power surges (intermittent, excessive outputs) from the power supply can also cause the transistor to burn out. In fact, power supply surges are a common cause of transistor burnout.

All of these conditions should be checked before placing a new transistor in the circuit. Some other malfunctions, along with their common causes, include

1. Burned-out transistors caused by thermal runaway. An increase in transistor current heats the transistor, causing a further increase in current, resulting in more heat. This continues until the heat dissipation capabilities of the transistor are exceeded. Bias-stabilization circuits are generally included in most well-designed transistor equipment.
2. Power supply overload caused by a short circuit in some portion of the voltage distribution network.
3. Burned-out transformer in shunt feed system caused by shorted blocking capacitor.
4. Blown fuses caused by power supply surges or shorts in filtering (power) networks.

It is impractical to list all of the common malfunctions and their related causes that you may find in troubleshooting electronic equipment. Generally, when a component fails, the cause is an operating condition which exceeded the maximum ratings of the component. However, it is possible for a component to simply "go bad."

The operating condition that causes a failure can be temporary and accidental, or it can be a basic design problem. Either way, your job is to find the trouble, verify its source or cause, and then repair it.

1–5.8. Repairing Troubles

In the strict sense, repairing the trouble is not part of the troubleshooting procedure. However, repair is an important part of the total effort involved in getting equipment back into operation. Repairs must be made before the equipment can be checked out and ready for operation.

Never replace a component if it fails a second time without first making sure that the cause of trouble is eliminated. Actually, the cause of trouble should be pinpointed before replacing a component the first time. However, this is not always practical. For example, if a resistor burns out because of an intermittent short and you have cleared the short, the next logical step is to replace the resistor. However, the short could happen again, burning out the replacement resistor. If so, you must recheck every element and lead in the circuit.

When replacing a defective component, an *exact replacement* should be used if it is available. If not available and if the original component is beyond repair, an equivalent *or better* component should be used. *Never* install a replacement component having characteristics or ratings inferior to those of the orginal.

An exception to this rule is when an equivalent or better component is not available and it is imperative that the equipment be placed in operation in the shortest possible time. Of course, when the emergency is over, the equivalent or better component must be installed when it becomes available.

Another factor to consider when repairing the trouble is that, if at all possible, the replacement component should be installed in the *same physical location as the original,* with the same lead lengths, and so on. This precaution is optional in most low-frequency or d-c circuits, but it must be followed for high-frequency applications.

In high-frequency (RF, IF, video, etc.) circuits, changing the location of components or the length of leads may cause the circuit to become detuned or otherwise out of alignment.

1-5.9. Operational Checkout

Even after the trouble is found and the faulty component is located and replaced, the troubleshooting effort is not necessarily completed. An operational check must be performed to verify that the equipment is *free of all faults* and is performing properly again. Never assume that simply because a defective component is located and replaced the equipment will automatically operate normally again. *In practical troubleshooting never assume anything; prove it!*

Operate the equipment through its *complete operating sequence.* This will ensure that one fault has not caused another. Follow the technical manual or data sheet procedures when available. On specialized equipment, have the operator go through the entire sequence but verify operation yourself.

When the operational check is completed and the equipment is again certified to be operating normally, make a brief record of the symptoms, faulty component, and remedy. This is particularly helpful when you must service the same equipment on a regular basis or when you must troubleshoot similar equipment. Even a simple record of troubleshooting will give you a history of the equipment for future reference.

If the equipment does not perform properly during the operational checkout, you must continue troubleshooting. If the symptoms are the same as (or similar to) the original trouble symptoms, retrace your steps one at a time. If the symptoms are entirely different, you may have to repeat the entire troubleshooting procedure from the start. However, this is usually not necessary.

For example, assume that the equipment does not check out because a replacement transistor has detuned the circuit (say, an IF amplifier transistor). When this is the case, you can repair the trouble by IF alignment rather than returning to the first step of troubleshooting and repeating the entire procedure. Keep in mind that you have arrived at the defective circuit or component by a systematic procedure. Therefore, retracing your steps—one at a time—is the logical method.

2. BASIC SOLID-STATE TROUBLESHOOTING

In chapter 1 we discussed basic troubleshooting techniques that can be applied to all types of electronic equipment—vacuum-tube, solid-state, sealed-module, integrated-circuit, and so on.

In this chapter we shall cover those troubleshooting techniques that apply primarily to solid-state (including IC) equipment. Although the basic principles of troubleshooting are used for solid-state, there are many practical techniques that apply to solid-state equipment only.

As one example, the key to most two-junction transistor troubleshooting is bias of the base-emitter junction. This junction must be forward-biased before the transistor will perform its function whether it is amplification, oscillation, wave-shaping, or whatever. Therefore, operation of a transistor and its related circuit can be checked by measurement of the forward bias or by temporary removal (or application) of forward bias.

These measurement techniques, together with many other test and trouble-localization methods for solid-state equipment, are discussed in the following sections. Specialized repair methods required for solid-state and IC equipment are also discussed.

2–1. MEASURING TRANSISTOR VOLTAGES IN CIRCUIT

It is possible to locate many defects in transistor circuits by measuring and analyzing the voltages at the base, emitter, and collector. This can be done with the circuit operating and without disconnecting any parts. Once located, the defective part can be disconnected and tested or substituted, whichever is most convenient.

Transistor circuits can be analyzed with an electronic voltmeter or with a very sensitive VOM. A number of manufacturers produce VOMs designed specifically for transistor troubleshooting. (The Simpson Model 250 is a typical example.) These VOMs have very low voltage scales to measure the differences that often exist between elements of a transistor (especially the small voltage difference between emitter and base). Such VOMs also have a very low voltage drop (50 mV) in the current ranges.

Figure 2–1 shows the basic connections for both PNP and NPN transistor circuits. The coupling or bypass capacitors have been omitted to simplify the explanation. The purpose of Fig. 2–1 is to establish *normal* transistor voltages. With a normal pattern established, it is relatively simple to find an abnormal condition.

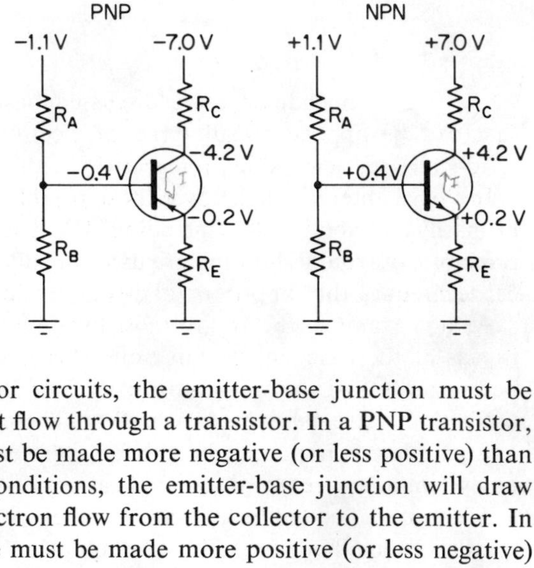

Figure 2–1. Basic connections for PNP and NPN transistor circuits (with normal voltage relationships).

In practically all transistor circuits, the emitter-base junction must be forward-biased to get current flow through a transistor. In a PNP transistor, this means that the base must be made more negative (or less positive) than the emitter. Under these conditions, the emitter-base junction will draw current and cause heavy electron flow from the collector to the emitter. In an NPN transistor, the base must be made more positive (or less negative) than the emitter for current to flow from emitter to collector.

2–1.1. Practical Analysis of Transistor Voltages

The following general rules are helpful in practical analysis of transistor voltages:

1. The middle letter in PNP or NPN always applies to the *base*.
2. The first two letters in PNP and NPN refer to the *relative* bias polarities of the *emitter* with respect to either the base or collector. For example, the letters PN (in *PNP*) indicate that the emitter is positive with respect to both the base and collector. The letters NP (*NPN*) indicate that the emitter is

negative with respect to both the base and collector.

3. The collector-base junction is always reverse-biased.
4. The emitter-base junction is usually forward-biased. An exception is a class C amplifier (used in RF circuits).
5. A *base input* voltage that opposes or decreases the forward bias also decreases the emitter and collector currents.
6. A *base input* voltage that aids or increases the forward bias also increases the emitter and collector currents.
7. The d-c electron flow is always against the direction of the arrow on the emitter.
8. If electron flow is into the emitter, electron flow will be out from the collector.
9. If electron flow is out from the emitter, electron flow will be into the collector.

Using these rules, normal transistor voltages can be summed up this way:

1. For a PNP transistor, the base is negative, the emitter is not quite as negative, and the collector is far more negative.
2. For an NPN transistor, the base is positive, the emitter is not quite as positive, and the collector is far more positive.

2-1.2. Practical Measurement of Transistor Voltages

There are two schools of thought on how to measure transistor voltages.

Some technicians prefer to measure transistor voltages from element to element (between electrodes) and note the *difference in voltage*. For example, in the circuit of Fig. 2-1, a 0.2-V differential would exist between base and emitter. The element-to-element method of measuring transistor voltages will quickly establish foward and reverse bias.

The most common method of measuring transistor voltages is to measure from a common or ground to the element. Service literature generally specifies transistor voltages this way. For example, all of the voltages for the PNP of Fig. 2-1 are negative with respect to ground. (The positive test lead of the meter must be connected to ground, and the negative test lead is connected to the elements, in turn.)

This method of labeling transistor voltages is sometimes confusing to those not familiar with transistors, since it appears to break the rules. (In a PNP transistor, both the emitter and collector should be positive, yet all of the elements are negative.) However, the rules still apply.

In the case of the PNP in Fig. 2-1, the emitter is at -0.2 V, whereas the base is at -0.4 V. The base is *more negative* than the emitter. Therefore, the emitter is *positive with respect to the base,* and the base-emitter junction is forward biased (normal).

On the other hand, the base is at -0.4 V, whereas the collector is at -4.2 V. The base is *less negative* than the collector. Therefore, the base is *positive with respect to the collector,* and the base-collector junction is reverse-biased (normal).

2-2. TROUBLESHOOTING WITH TRANSISTOR VOLTAGES

The following is an example of how voltages measured at the elements of a transistor can be used to analyze failure in solid-state circuits.

Assume that a PNP transistor circuit is measured and the voltages found are similar to those of Fig. 2-2. Except in one case, these voltages indicate a defect. It is obvious that the transistor is not forward-biased because the emitter is more negative than the base (reverse bias for a PNP). The only circuit where this might be normal is one that requires a large *trigger* voltage or pulse (negative in this case) to turn it on. Such circuits are found in RF amplifiers and some switching applications.

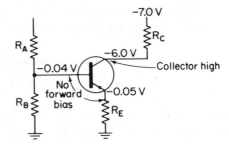

Figure 2-2. PNP transistor circuit with abnormal voltages (emitter-base not forward biased, collector voltage high).

The first clue of an error in Fig. 2-2 is that the collector voltage is almost as large as the collector source (at R_C). This means that very little current is flowing through R_C in the collector-emitter circuit. The transistor could be defective. But the trouble is more likely caused by a problem in bias. The emitter voltage depends mostly on the current through R_E so unless the value of R_E has changed substantially (this would be unusual), the problem is one of incorrect bias on the base.

The next step in this case is to measure the bias-source voltage at R_A. If the bias-source voltage is, as shown in Fig. 2-3, at -0.9 V instead of the required -1.1 V, the problem is obvious; the external bias voltage is incorrect. This condition will probably show up as a defect in the power supply and will appear as an incorrect voltage in other circuits.

If the source voltage is correct, as shown in Fig. 2-4, the cause of trouble is probably a defective R_A or R_B or a defect in the transistor.

The next step is to remove all voltage from the equipment and measure the resistance of R_A and R_B. If either value is incorrect, the corresponding resistor must be replaced. If both values are correct, it is reasonable to check

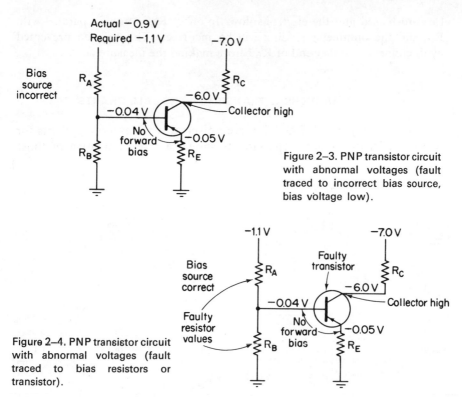

Figure 2–3. PNP transistor circuit with abnormal voltages (fault traced to incorrect bias source, bias voltage low).

Figure 2–4. PNP transistor circuit with abnormal voltages (fault traced to bias resistors or transistor).

the value of R_E. However, it is more likely that the transistor is defective. This can be established by test and/or replacement.

2–2.1. Practical In Circuit Resistance Measurements

Do not attempt to measure resistance values in transistor circuits with the resistors still connected. This practice may be correct for vacuum-tube circuits but not with transistor circuits. One reason is that the voltage produced by the ohmmeter battery could damage some transistors. Even if the voltages are not dangerous, the chance for an error is greater with a transistor circuit because the transistor junctions will pass current in one direction. This could complete a circuit through other resistors and produce a series or parallel combination, thus making false indications. *This can be prevented by disconnecting one resistor lead before making the resistance measurement.*

For example, assume that an ohmmeter is connected across R_B (Figs. 2–2 through 2–4) with the positive battery terminal of the ohmmeter connected to ground. Because R_E is also connected to ground, the positive battery terminal is connected to the end of R_E. Since the battery negative terminal is connected to the transistor base, the emitter-base junction is forward-biased and the electrons flow. In effect R_E is now in parallel with R_B, and the ohmmeter reading will be incorrect. This can be prevented by disconnecting either end of R_B before making the measurement.

2–3. UNIVERSAL TRANSISTOR TROUBLE CHARTS

Figures 2–5 and 2–6 are universal troubleshooting charts for typical transistor circuits. The circuits shown are representative of those

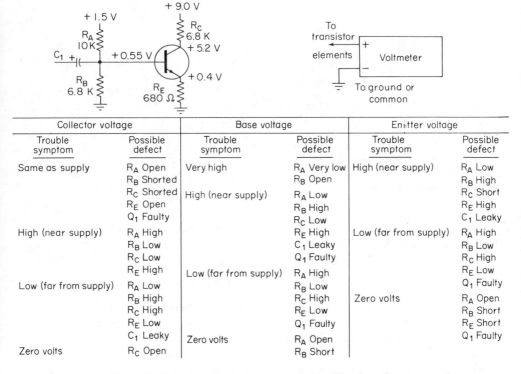

Collector voltage		Base voltage		Emitter voltage	
Trouble symptom	Possible defect	Trouble symptom	Possible defect	Trouble symptom	Possible defect
Same as supply	R_A Open R_B Shorted R_C Shorted R_E Open Q_1 Faulty	Very high	R_A Very low R_B Open	High (near supply)	R_A Low R_B High R_C Short R_E High C_1 Leaky
		High (near supply)	R_A Low R_B High R_C Low		
High (near supply)	R_A High R_B Low R_C Low R_E High		R_E High C_1 Leaky Q_1 Faulty	Low (far from supply)	R_A High R_B Low R_C High R_E Low Q_1 Faulty
		Low (far from supply)	R_A High R_B Low R_C High R_E Low Q_1 Faulty		
Low (far from supply)	R_A Low R_B High R_C High R_E Low C_1 Leaky			Zero volts	R_A Open R_B Short R_E Short Q_1 Faulty
Zero volts	R_C Open	Zero volts	R_A Open R_B Short		

Figure 2–5. Universal transistor troubleshooting chart (two sources).

found in radio and TV receivers, audio equipment, and a number of commercial and industrial applications.

Figure 2–5 shows an NPN transistor circuit where the base bias network is supplied by a source different from the collector. Figure 2–6 shows an NPN transistor supplied from a single negative source. That is, the collector and bias networks are returned to ground or common. The troubleshooting information in Figs. 2–5 and 2–6 will also hold for PNP transistors except

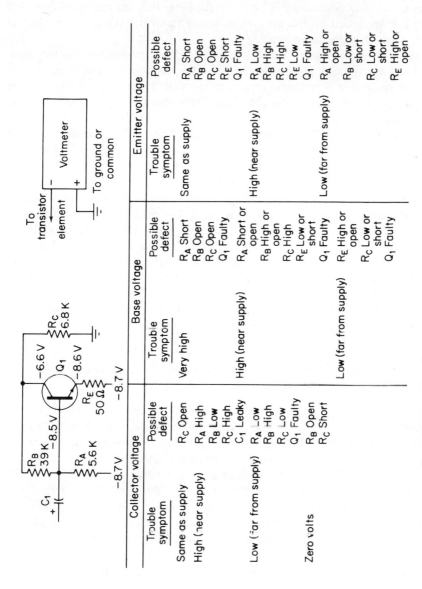

Figure 2–6. Universal transistor troubleshooting chart (negative supply).

that the supply polarities will be reversed. Note that the charts of Figs. 2–5 and 2–6 are based on trouble symptoms of abnormal voltages measured at the collector, base, and emitter.

The first step in using Figs. 2–5 and 2–6 is to measure the collector, base, and emitter voltages *in reference to ground or common* on the circuit board or chassis, rather than from element to element. Keep in mind that voltage deviations might be very slight. Therefore, the voltages should be read carefully.

Compare the voltage deviations (if any) with those in the trouble symptom columns. Match up the possible defects for *all three elements* (collector, base, and emitter) crossing off those that do not match. The remaining possible defects then become the most likely defects.

For example in Fig. 2–5 assume that the collector voltage is high with base and emitter voltages low. Resistors R_E and R_C could be crossed off as most likely defective components. A high value of R_E would match the high collector voltage but not the low emitter and base voltage symptoms. The same is true for a low value of R_C. Resistors R_A and R_B are most likely defective components, since a high R_A and a low R_B would produce all three symptoms.

When none of the symptoms appear to match, suspect the transistor or possibly the coupling capacitor rather than the resistors.

2–4. TESTING TRANSISTORS IN CIRCUIT (FORWARD-BIAS METHOD)

Germanium transistors normally have a 0.2-to 0.4-V *voltage differential* between emitter and base; silicon transistors normally have a voltage differential of 0.4 to 0.8 V. The polarities of voltages at the emitter and base will depend on the type of transistor (NPN or PNP).

The voltage differential between emitter and base acts as a forward bias for the transistor. That is, a sufficient voltage differential or forward bias will turn the transistor on, resulting in a corresponding amount of emitter-collector current flow. Removal of the voltage differential or an insufficient differential will produce the opposite results. That is, the transistor will be cut off (no emitter-collector flow, or very little flow).

These forward-bias characteristics can be used to test transistors in circuit, without an in-circuit tester. The following sections describe two methods of testing transistors in circuit: one by removing the forward bias and the other by introducing a forward bias.

2–4.1. Removal of Forward Bias

Figure 2–7 shows the test connections for an in-circuit transistor test by removal of forward bias. The procedure is simple. First, measure

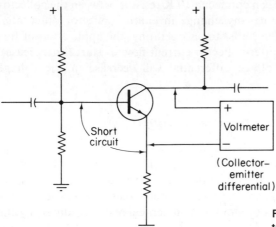

Figure 2–7. In-circuit transistor test (removal of forward bias).

the emitter-collector differential under normal circuit conditions. Then short the emitter-base junction and note any change in emitter-collector differential. If the transistor is operating, the removal of forward bias will cause the emitter-collector current flow to stop, and the emitter-collector voltage differential will increase. That is, the collector voltage will rise to or near the source value.

2–4.2. Application of Forward Bias

Figure 2 8 shows the test connections for an in-circuit transistor test by the application of forward bias. The procedure is equally simple. First, measure the emitter-collector differential (or voltage across R_E) under

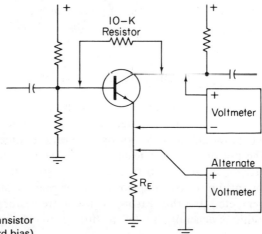

Figure 2–8. In-circuit transistor test (application of forward bias).

normal circuit conditions. Then connect a 10 K resistor between the collector and base, as shown, and note any change in emitter-collector differential (or voltage across R_E). If the transistor is operating, the application of forward bias will cause the emitter-collector current flow to start (or increase), and the emitter-collector voltage differential will decrease or the voltage across R_E will increase.

2–4.3. Go/No Go Test Characteristics

The test methods shown in Figs. 2–7 and 2–8 demonstrate that the transistor is operating on a go/no go basis. This is usually sufficient for most d-c and low-frequency a-c applications. However, the tests *do not* show transistor gain or leakage. Also, the tests do not establish operation of the transistor at high frequencies or show how much delay is introduced by the transistor.

The fact that these (or similar) in-circuit tests of a transistor will not establish all of the operating characteristics, particularly those that affect high-frequency RF or rapid-pulse switching applications, raises a problem.

Some technicians reason that the only satisfactory test of a transistor is in-circuit operation. If a transistor will not perform its function in a given circuit, the transistor must be replaced. Therefore, the most logical method of test is replacement.

This reasoning is generally sound, except in one circumstance. It is possible that a replacement transistor will not perform satisfactorily in critical circuits (high-frequency RF and high-speed switching circuits), even though the transistor is the correct type and may work in another circuit. This can be misleading. If such a replacement transistor does not restore the circuit to normal, the apparent fault is with another circuit part, whereas the true cause of trouble is the new transistor. Fortunately, this does not happen often, even in critical circuits. However, a good troubleshooter should be aware of the possibility.

2–5. TRANSISTOR TESTERS

Transistors can be tested in or out of circuit using commercial transistor testers. These testers are the solid-state equivalent of vacuum-tube testers (although they do not operate on the same principle).

The use of transistor testers in solid-state troubleshooting is generally a matter of opinion. At best, such testers show the gain and leakage of transistors at d-c or low frequencies under one set of conditions (fixed voltage, current, etc.). For this reason, the use of transistor testers in practical troubleshooting is generally limited to home-entertainment-type equipment. Transistors used in high-frequency or switching applications are tested by substitution (in circuit) or with special test equipment (out of circuit). The procedures and devices for an out-of-circuit test, such as curve tracers, measurement of switching time, and so on, are discussed in the author's *Practical Semiconductor Databook for Electronic Engineers and Technicians*, Prentice-Hall Inc., Englewood Cliffs, N.J., 1969.

2–5.1. Typical Commercial Transistor Tester

It is impractical to discuss every type of transistor tester circuit available. One very effective method for both in-circuit and out-of-circuit tests is shown in Fig. 2–9. With this method, a 60-Hz squarewave pulse is applied to both the base-emitter and collector-emitter junctions simultaneously. The current flow in each of the two junctions is measured and compared. The difference in current flow (collector-emitter divided by base-emitter) is the transistor gain.

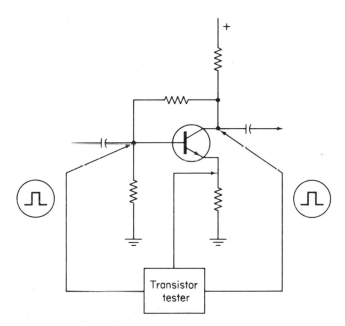

Figure 2–9. In-circuit transistor test method.

It is not necessary to remove the transistor from the circuit to make such a test although the equipment must be turned off. In fact, this test method can also show up defects in the transistor circuit.

For example, in many transistor circuits, the overall gain is set by circuit resistance values rather than transistor gain. Assume that a particular circuit has resistance values that would normally show a gain of 10, and the in-circuit squarewave test (using a tester) shows no gain or very low gain. This could be the result of a bad transistor or circuit problems. If the transistor is then tested out of circuit under identical conditions and a gain is shown, the problem is most likely one of an undesired change in circuit resistance values.

2–6. TESTING TRANSISTORS OUT OF CIRCUIT

There are four basic tests required for transistors in practical troubleshooting: gain, leakage, breakdown, and switching time. All of these tests are best made with an oscilloscope using appropriate adapters (curve tracers, switching-characteristic checkers, etc.). However, it is possible to

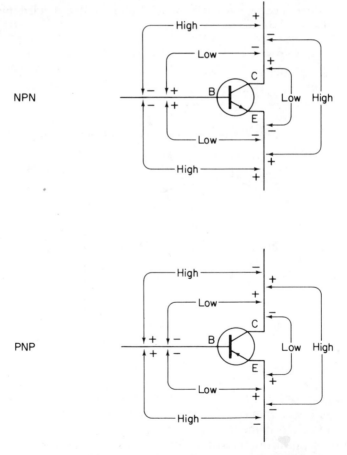

Figure 2–10. Transistor leakage tests with an ohmmeter.

test a transistor with an ohmmeter. These simple ohmmeter tests will show if the transistor has leakage and if the transistor shows some gain.

As discussed, the only true test of a transistor is in the circuit with which the transistor is to be used.

2–6.1. Testing Transistor Leakage with an Ohmmeter

For meter test purposes, transistors can be considered as two diodes connected back to back. Therefore, each diode should show low forward resistance and high reverse resistance. These resistances can be measured with an ohmmeter as shown in Fig. 2–10.

The same ohmmeter range should be used for each pair of measurements (base-to-emitter, base-to-collector, and collector-to-emitter). However, avoid using the R × 1 range or an ohmmeter with a high internal battery voltage. Either of these conditions can damage a low-power transistor.

If the reverse resistance reading is low but not shorted, the transistor is leaking. If both forward and reverse readings are very low or show a short, the transistor is shorted. If both forward and reverse readings are very high, the transistor is open. If the forward and reverse readings are the same or nearly equal, the transistor is defective.

A typical forward resistance is 300 to 700 Ω. However, a low-power transistor might show only a few ohms in the forward direction especially at the collector-emitter junction. Typical reverse resistances are 10 to 60 K.

Actual resistance values will depend on the ohmmeter range and battery voltage. Therefore, the *ratio of forward-to-reverse* resistance is the best indicator. Almost any transistor will show a ratio of at least 30 to 1. Many transistors show ratios of 100 to 1 or greater.

2–6.2. Testing Transistor Gain with an Ohmmeter

Normally, there will be little or no current flow between emitter and collector until the base-emitter junction is forward-biased. Therefore, a basic gain test of a transistor can be made using an ohmmeter. The test circuit is shown in Fig. 2–11. In this test the R × 1 range should be used. Any internal battery voltage can be used provided that it does not exceed the maximum collector-emitter breakdown voltage.

In position A of switch S_1, there is no voltage applied to the base, and the base-emitter junction is not forward-biased. Therefore, the ohmmeter should read a high resistance. When switch S_1 is set to B, the base-emitter circuit is forward-biased (by the voltage across R_1 and R_2) and current flows in the emitter-collector circuit. This is indicated by a lower resistance reading on the ohmmeter. A 10-to-1 resistance ratio is typical for an AF transistor.

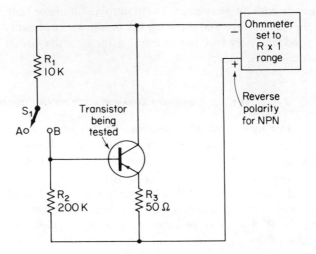

Figure 2–11. Transistor gain test with an ohmmeter.

2–7. TESTING DIODES OUT OF CIRCUIT

Three basic tests are required for diodes. First, the diode must have the ability to pass current in one direction (forward current) and prevent or limit current flow (reverse current) in the opposite direction. Second, for a given reverse voltage, the reverse current should not exceed a given value. Third, for a given forward current, the voltage drop across the diode should not exceed a given value. If a diode is to be used in pulse or digital work, the

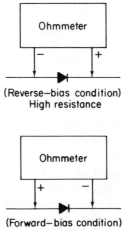

Figure 2–12. Basic diode test with an ohmmeter.

switching time must also be tested. These tests are best performed using an oscilloscope with appropriate adapters.

However, since the elementary purpose of a diode is to prevent current flow in one direction while passing current in the opposite direction, a diode can be tested using an ohmmeter. In this case the ohmmeter is used to measure forward and reverse resistance of the diode. The basic circuit is shown in Fig. 2–12.

A good diode will show high resistance in the reverse direction and low resistance in the forward direction.

If resistance is low in the reverse direction, the diode is probably leaking. If resistance is high in both directions, the diode is probably open. A low resistance in both directions usually indicates a shorted diode. It is possible for a defective diode to show a difference in forward and reverse resistance. The important factor in making a diode-resistance test is the *ratio of forward-to-reverse resistance* (often known as the front-to-back ratio or the back-to-front ratio). The actual ratio will depend upon the type of diode. However, as a rule of thumb, a small signal diode will have a ratio of several hundred to one, while a power rectifier can operate satisfactorily with a ratio of 10 to 1.

2–8. TESTING MISCELLANEOUS SOLID-STATE COMPONENTS

Solid-state components such as unijunction transistors (UJTs), field-effect transistors (FETs), voltage-variable diodes (VVCs), and silicon-controlled rectifiers (SCRs) are best tested by monitoring the input and output waveforms during in-circuit operation. This is done as part of the troubleshooting procedure. The waveforms are compared against those in the service literature.

Such solid-state devices can also be tested out of circuit using an oscilloscope with appropriate adapters. There are no satisfactory quick tests for most solid-state components. That is, the quick tests prove very little from a practical troubleshooting standpoint. Generally, substitution is the best and ultimate test.

2–8.1. Handling IGFETs

If it becomes necessary to remove an insulated-gate field-effect transistor (IGFET) for test (by substitution or on an external tester) during the troubleshooting sequence, the IGFET *must be* treated with care.

The IGFET, often known as a *metal oxide semiconductor field-effect transistor,* or MOSFET, due to its physical structure, is a very delicate device out of circuit. In circuit, the IGFET is just as rugged as the junction-type field-effect transistor, or JFET. Out of circuit, the IGFET is subject to damage from static charges when handled. The IGFET is generally shipped with the leads all shorted together to prevent damage in shipping and handling (there will be no static discharge between leads).

To test an IGFET out of circuit, the leads from the tester should be connected *before the short is removed* from the IGFET leads. The following points should be considered when checking an IGFET (or MOSFET).

1. First, turn the power off. When the IGFET is to be removed from a unit for testing, your body should be at the same potential as the unit. This can be done by placing one hand on the chassis before removing the IGFET. Before the IGFET is connected to an external test circuit, put the hand holding the IGFET against the tester front panel and connect the source lead from the tester to the source lead of the IGFET. This procedure prevents possible damage from static charges to the IGFET. If you must test several IFGETs during the troubleshooting procedure, clip a lead from the tester to your watch band or ring and touch the chassis before removing the IGFETs.

   Warning. Make certain that the chassis is at ground potential before touching it. In some obsolete or defective equipment, the chassis or panel is above ground (typically by one-half the line voltage).

2. When handling an IGFET, the leads must be shorted together. Generally, this is done in shipment by a shorting ring a piece of wire or a piece of foil. Connect the tester leads (source lead first) to the IGFET. Then remove the shorting ring wire or foil.

3. When soldering or unsoldering an IGFET, the soldering tool tip must be at ground potential (no static charge). Connect a clip lead from the barrel of the soldering tool to the tester case. The use of a soldering gun is not recommended for IGFETS.

4. Remove power to the circuit before inserting or removing an IGFET (or a plug-in module containing an IGFET). The voltage transients developed when terminals are separated may damage the IGFET. This same caution applies to circuits with conventional transistors. However, the chances of damage are greater with IGFETs, due to their delicate nature.

2–9. TESTING ICs

There is some difference of opinion on testing ICs in circuit or out of circuit during troubleshooting. An in-circuit test is the most convenient, since the power source is available and you do not have to unsolder the IC. (This can be quite a job, as discussed in sec. 2–10. 2.)

Of course, first you must measure the d-c voltages applied at the IC terminals to make sure that they are available and correct. If the voltages are absent or abnormal, this is a good starting point for troubleshooting.

With the power sources established, the in-circuit IC is tested by applying the appropriate input and monitoring the output. In some cases it is not necessary to inject an input, since the normal input will be supplied by the circuits ahead of the IC.

One drawback to testing an IC in circuit is that the circuits before (input) and after (output) the IC may be defective. This could lead you to think that the IC was bad. For example, assume that the IC is an AND gate, requiring two inputs to produce an output. (AND gates are described fully in Chapter 3.) To test such an IC you apply two input pulses, and monitor the IC for an output pulse. Now assume that the IC output terminal is connected to a short circuit. There will be no output indicated, even though the IC and the two input pulses are good. Of course, this will show up as an incorrect resistance measurement (if such measurements are made).

Out-of-circuit tests for ICs have two obvious disadvantages: you must remove the IC and you must supply the required power. However, if you test a suspected IC after removal and find that it is operating properly out of circuit, it is logical to assume that there is trouble in the circuits connected to the IC. This is very convenient to know *before* you go to the trouble of installing a replacement IC.

2–9.1. IC Voltage Measurements

While the operating test procedures for an IC are the same as for conventional transistor circuits of the same type, the measurement of static (d-c) voltage applied to the IC is not identical. Most ICs require connection to both a positive and negative power source. A few ICs can be operated from a single power supply source.

Many ICs require equal power supply voltages (such as +9 V and –9 V). However, this is not the case with the example circuit of Fig. 2–13, which requires +9 V at pin 8 and –4.2 V at pin 4.

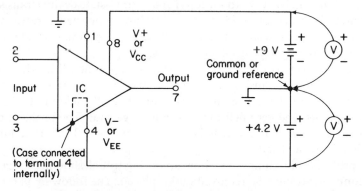

Figure 2–13. Measuring static (power source) voltages of ICs.

Unlike most transistor circuits, where it is common to label one power supply lead positive and the other negative without specifying which (if either) is common or ground, it is *necessary that all IC power supply voltages be referenced to a common or ground.*

Manufacturers do not agree on power supply labeling for ICs. For example, one manufacturer might use V + to indicate the positive voltage and V− to indicate the negative voltage. Another manufacturer might use the symbols V_{EE} and V_{CC} to represent negative and positive, respectively. For this reason, the schematic diagram should be studied carefully before measuring power source voltages during troubleshooting.

No matter what labeling is used, the IC will usually require two power sources, with the positive lead of one and the negative lead of the other tied to ground. Each voltage must be measured separately, as shown in Fig. 2–13.

Note that the IC case (such as a TO-5 type) of the Fig. 2–13 circuit is connected to pin 4. Such a connection is typical for most ICs (but not necessarily pin 4). Therefore, the case will be below ground (or hot) by 4.2 V.

2–10. WORKING WITH TOOLS IN SOLID-STATE EQUIPMENT

It is assumed that the reader is familiar with common handtools used in electronics service, such as diagonal wire cutters, long-nose pliers, soldering tools (both pencil type and soldering guns), insulated probes, nut drivers, and wrenches. These same tools are used in solid-state equipment. However, certain additional tools and techniques are also required (or will make life much easier if used).

The main problem in working with solid-state equipment is *heat*. Typically, solid-state devices are made of germanium or silicon. Neither of these junction materials can withstand high temperatures for any length of time. A temperature of 200 °C is about tops for any solid-state device (transistor or IC). Since the leads of solid-state devices are connected directly to the junction material, the leads can not be subjected to continuous and excessive heat when soldering and unsoldering leads.

Another problem is the *size* of solid-state components. With the possible exception of power transistors, the cases and leads of transistors can be considered as miniature. This means the use of small soldering tools, pliers with fine points, and very delicate handling in general. The problem of size is equally great with ICs which have many fine wire leads that must be unsoldered and soldered.

The problems of heat and size show up during troubleshooting when solid-state components must be removed and replaced. The following paragraphs describe practical techniques for working with solid-state equipment.

2–10.1. Working with Printed Circuit Components

Solid-state components are often mounted on printed circuit or etched circuit boards. In turn, the boards may be bolted to a main chassis or plugged into a connector on the chassis. Either way, the component leads must pass through eyelets (sometimes known as soldering cups or, simply, holes) in the board. The eyelets make contact with the printed or etched wiring. The component leads must be soldered to the eyelets as shown in Fig. 2–14.

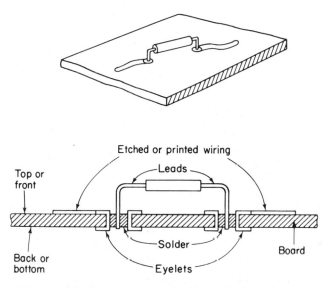

Figure 2–14. Component mounted on printed circuit board.

When executed properly, removal and replacement of components on a printed board is not impossible, although it may be difficult. The following techniques are recommended to replace components on etched or printed boards. The techniques will be satisfactory for most practical applications. Of course, you must modify the procedure as necessary to fit the particular equipment.

2–10.1.1. Basic Printed Circuit Rework Procedure

It is best to remove the board from the equipment unless the back of the board is accessible. Unless specified otherwise in the service literature, use electronic grade 60/40 solder and a 15-W pencil soldering tool. A higher-wattage soldering tool, if applied for too long a period, can ruin the bond between the etched wiring and the insulating base material by charring the glass epoxy laminate or whatever material is used for the base. However, a 40-to 50-W soldering tool can be used if the touch-and-wipe technique is followed, as described later in this section. The author prefers a chisel tip for the soldering tool, about 1/16 in wide. However, this is a matter of choice.

If the component is to be removed and replaced with a new part, cut the leads near the body of the component as shown in Fig. 2–15. This will free the leads for individual unsoldering. Grip the lead with long-nose pliers. Apply the tip of the soldering tool to the connection *at the back* of the board; then pull gently to remove the lead.

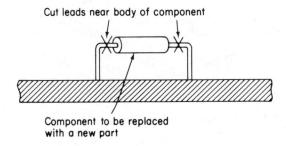

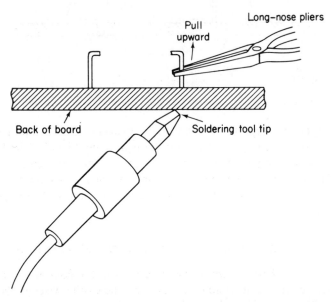

Figure 2–15. Basic printed circuit rework procedure (when component is to be replaced).

If the component is to be removed for test and possibly reinstalled, do not cut the leads. Instead, grip the lead from the front with long-nose pliers and apply the soldering tool to the connection at the back of the board. Lift the lead straight out as shown in Fig. 2–16.

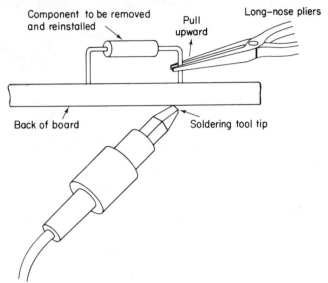

Figure 2–16. Basic printed circuit rework procedure (when component is to be removed and reinstalled).

Note that the pliers will provide a heat sink for the component being removed. This will keep the component from heating if it is necessary to apply the soldering tool for a long time (which should be avoided). If a transistor is to be unsoldered, an alligator clip attached to the transistor case as shown in Fig. 2–17 will provide a temporary heat sink. Soldering heat sinks are also available commercially.

When the lead comes out of the board, the lead should leave a clean hole. If it does not, the hole should be cleaned by reheating the solder and placing a sharp object such as a toothpick or enameled wire into the hole to clean out the old solder. Some technicians prefer to blow out the hole with compressed air. This is *never* recommended when the board is in or near the equipment, since the solder spray could short other circuits on other parts of the equipment.

An alternative method of removing solder is with a *solder gobbler* such as shown in Fig. 2–18. There are many versions of such solder gobblers (also known as vacuum desolderers). The tool shown in Fig. 2–18 is hand operated and is used mostly for repair work. There are vacuum-pump-operated models available, but they are used mostly in assembly-line work.

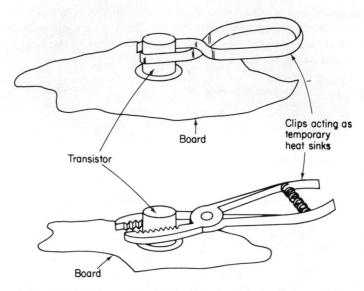

Figure 2–17. Using clips as transistor heat sinks during soldering and desoldering.

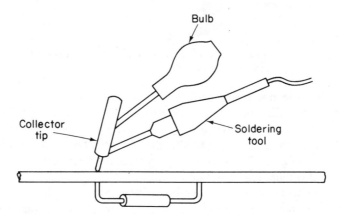

Figure 2–18. Typical hand-operated desoldering tool (solder gobbler).

As shown in Fig. 2–18, the gobbler tool consists of a soldering tool, collector tip, and bulb. In use, the bulb is squeezed, the collector tip is placed on the solder area, and when the solder is molten, the bulb is released, drawing the solder into the collector tip. Then the solder can be forced from the tip by squeezing the bulb again. The tool can also be used for soldering. Solder gobblers are especially effective for removing ICs from a circuit, as is discussed later in this section.

Some manufacturers do not recommend either a solder gobbler or the conventional desoldering technique. Their argument is that either method requires prolonged heating. As an alternative, they recommend a touch-and-wipe technique. With this method, a hair-bristle soldering brush is used which has had the bristles shortened to about $\frac{1}{4}$ in. in length. A conventional soldering tool of 50 W or less is used. The tip of the tool is touched to the area where the component lead comes through the board with the brush ready to wipe the melted solder away. The solder is removed by a series of touch-and-wipe operations. The tool is touched to the board only long enough to melt the solder and is removed as the brush is wiped across the joint. After a few touch-and-wipe operations, the component lead can be bent away from the etched connection and removed.

Once the component lead is removed and the hole is clear, clean the leads on the new component and bend them to the correct shape to fit into the holes. Insert the leads, making certain the component seats exactly the same as the original. If it does not, reheat the connection and gently press the component into place.

Apply the soldering tool to the connection at the back of the board and apply only the amount of solder required to form a good electrical connection. In all cases the tip of the tool should be clean and properly tinned for best heat transfer in a short time to a soldered connection.

Check the front or component side of the board to make sure that the solder has filled through the plated hole. In some cases, the component may require soldering to an etched lead on the top (or front) of the board. In this case, apply the tip of the soldering tool to the connection and apply the required amount of solder to form a good electrical connection.

If it is necessary to hold a bare wire in place while soldering, a handy tool for this purpose can be made by cutting a notch into one end of a wooden tool as shown in Fig. 2–19. There are commercial versions of such wire-holding tools. Usually, the commercial soldering aids are made of chrome-plated steel. (The solder will not stick to the chrome plating.)

Clip the excess lead that protrudes through the hole in the board. Be sure that these ends are not dropped into the equipment where they could cause

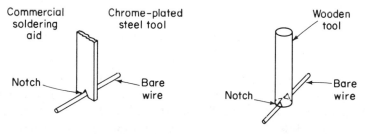

Figure 2–19. Typical wire-holding tools.

a short circuit. If necessary, clean the area around the soldered connection with a flux-remover solvent.

2-10.1.2. Handling Power Transistors

The foregoing procedures can be used for all components, including most transistors. Power transistors are a possible exception. Some power transistors are mounted on printed circuit boards and are provided with heat sinks (usually the fin type shown in Fig. 2–20). In other cases, power transistors are mounted on the chassis, with the chassis metal acting as the heat sink as shown in Fig. 2–21. Power transistors are sometimes mounted in Teflon sockets and may or may not be bolted to the chassis. While it may be necessary to remove heat sinks from transistors during repair, *never operate a power transistor without its heat sink*.

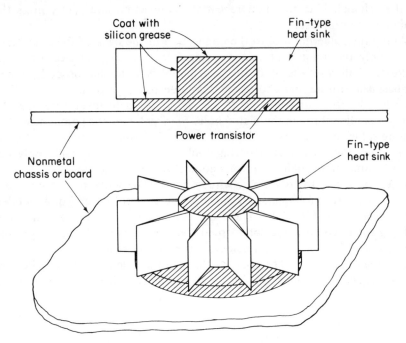

Figure 2–20. Typical mounting for power transistors on nonmetal chassis or boards (external heat sinks).

When replacing power transistors, be sure to use silicon grease to ensure maximum heat transfer from the transistor. Both sides of the mica insulation should be covered for best results. Be sure also that no foreign matter such as metal shavings stick to the mica insulator. This can cause a breakdown at a later time.

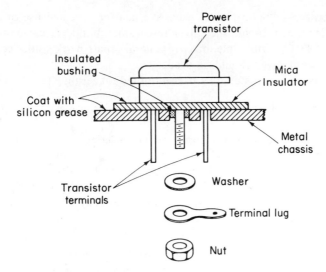

Figure 2–21. Typical mounting for power transistors on metal chassis (chassis acting as heat sink).

2–10.1.3. Strip-Soldering Techniques

Many solid-state components are mounted on strips, usually made of a ceramic material. Strip mounting is used instead of (or in addition to) printed circuit mounting, especially on equipment where the parts are not of the plug-in type.

Typical strip mounting is shown in Fig. 2–22. Often, the notches in these strips are lined with a silver alloy. This should be verified by reference to the

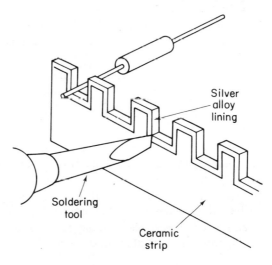

Figure 2–22. Strip-soldering techniques.

service literature. Application of excessive heat or repeated use of ordinary 60/40 tin/lead can break the silver-to-ceramic bond. Occasional use of ordinary solder is permissible, but for general repair work solder containing about 3 per cent silver should be used.

When removing or installing a part mounted on ceramic terminal strips.

1. Use a 50-W soldering tool. The tip should be tinned with silver-bearing solder.
2. Apply heat by touching one corner of the soldering tool tip to the base of the notch as shown in Fig. 2–22. Do not force the tip of the tool into the terminal notch. This may chip or break the ceramic strip.
3. Use the minimum amount of heat required to make the solder flow freely.
4. Apply only enough solder to form a good bond. Do not attempt to fill the notch with solder.

2–10.1.4. Soldering Large Metal Terminals

When soldering large metal terminals (switch terminals, potentiometer terminals, etc.) found in solid-state equipment, ordinary 60/40 solder is satisfactory. However, a larger soldering tool is required for such terminals. The 15-W soldering tool recommended for printed or etched circuits is usually too small for the large terminals. A 40- to 50-W soldering tool will do most jobs found in solid-state electronics. A soldering gun is also useful.

When soldering large terminals, use good electronic soldering practices; that is, apply only enough heat to make the solder flow freely and apply only enough solder to form a good electrical connection. Too much solder may impair operation of the circuit. An excess amount of solder can also cover a cold solder joint.

Clip off any excess wire that may extend past the solder connection. If necessary, clean the solder connection with flux-remover solvent.

2–10.2. Working with ICs

At first glance, it may appear that an IC cannot be removed from a printed board or socket without destroying or seriously damaging the IC. Any of the three IC packages in common use (flat-pack, TO-5-style case, or dual in-line, shown in Fig. 2–23) have many fine metal contacts or leads soldered in place. All leads must be unsoldered before the IC can be removed. This brings up obvious problems. You cannot use the standard unsoldering procedure discussed in foregoing paragraphs (pulling each lead with long-nose pliers while applying heat to the connection). That is, such a procedure is not possible in some cases (due to arrangement of the leads) and is highly impractical in most cases.

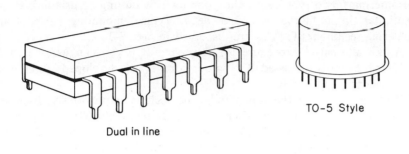

TO-5 Style

Dual in line

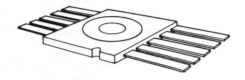

Flat pack

Figure 2–23. Typical IC packages.

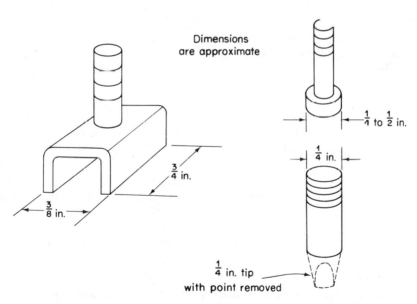

Dimensions
are approximate

$\frac{1}{4}$ to $\frac{1}{2}$ in.

$\frac{1}{4}$ in.

$\frac{3}{4}$ in.

$\frac{3}{8}$ in.

$\frac{1}{4}$ in. tip
with point removed

Figure 2–24. Typical desoldering tool tips for IC removal.

There are two practical solutions to the problem. First, you can use a *desoldering tool tip* that will contact all of the lead connections simultaneously. Such tips are shown in Fig. 2–24. There are commercial versions of IC desoldering tips or you can make your own by cutting or grinding a conventional tip to fit the particular need. Typical mounting patterns and soldering techniques for ICs are discussed in Sec. 2–10.2.1.

As an alternative procedure for removing ICs, you can use the *solder gobbler* previously discussed. The solder gobbler was developed originally as a desoldering tool for ICs.

You can also use the touch-and-wipe technique previously discussed. However, this is very tedious when many leads are involved (typically, a dual in-line IC package has 14 leads). Touch-and-wipe should be used only in emergencies, when special desoldering tips and solder gobblers are not available.

The main advantage to the special desoldering tip is speed. The main disadvantage is the excess heat which might damage the IC semiconductor material. When using a special tip that covers all contacts simultaneously, be ready to remove the IC immediately (just as soon as the solder is molten). It is not necessary to use the desoldering tips for soldering the IC back in place. This produces considerable unrequired heat.

The main advantage of the solder gobbler is the absence of excess heat. Also, the solder will be removed from the hole (usually). Of course, it takes more time to remove an IC with a solder gobbler than with a desoldering tip.

When installing a replacement IC, the mounting pattern (including lead bending, if any) of the original IC must be followed *exactly*. We say exactly since there is rarely enough space in IC equipment to do otherwise.

2–10.2.1. Typical IC Mounting Patterns and Techniques

To help the reader select the proper tools for IC work and to understand the practical problems of ICs, the following paragraphs describe typical IC installations.

Flat packs. Figure 2–25 shows five methods for making solder connections to flat packs. In the *straight-through method* (Fig. 2–25 a), the leads are bent downward for a 90° angle and are inserted in the circuit-board holes. When assembled at the factory, all leads are connected simultaneously by dip soldering or wave soldering. During repair, the leads are soldered one at a time. A disadvantage of the straight-through method is that the IC package must be held firmly in position during the soldering operation.

The *clinched-lead, full-pad soldering method* (Fig. 2–25b) requires an additional operation (clinching the lead) but has the advantage that the IC does not have to be held in position during soldering.

The *clinched-lead, offset-pad* (Fig. 2–25 c) and the *clinched-lead, half-pad* (Fig. 2–25 d) methods are variations of the clinched-lead, full-pad method.

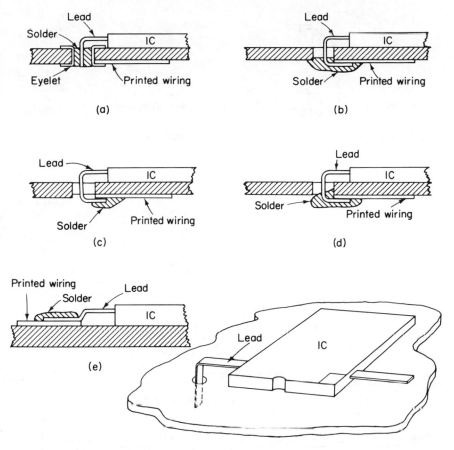

Figure 2–25. Typical soldering techniques for flat-pack ICs.

The half-pad and offset-pad methods are easier to replace since the hole is not filled with solder.

In the *surface-connection method* (Fig. 2–25 e), the connections are made on the package side of the board. This method is often used when ICs must be mounted on both sides of a board. No holes are required in the board. The *reflow soldering technique* is often used when surface-connection ICs are assembled at the factory. With reflow soldering both the surface contacts and the leads are tinned and covered with solder. Then the leads are set on the surface contact, and heat is applied to all contacts (or all contacts on one side). The heat causes the solder to reflow and make a good connection between leads and surface contacts.

Typical mounting patterns for flat-pack ICs are shown in Fig. 2–26. Each

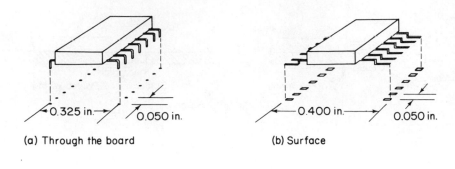

(a) Through the board   (b) Surface

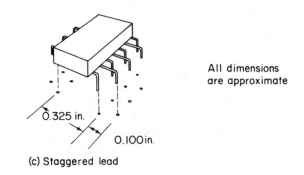

All dimensions
are approximate

(c) Staggered lead

Figure 2–26. Typical mounting patterns for flat-pack ICs.

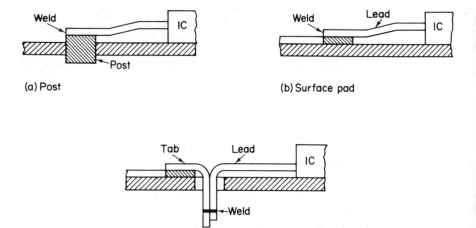

(a) Post   (b) Surface pad

(c) Tab

Figure 2–27. Typical welding techniques for flat-pack ICs.

of these patterns has advantages and disadvantages from a design standpoint. However, from a practical troubleshooting standpoint, the technician is not concerned with design problems. Instead the technician should study the existing mounting pattern to find the simplest and most effective method of removal and replacement.

In some cases, flat-pack ICs are attached to the circuit board by means of a weld rather than soldering. Such leads must be clipped when the IC package is to be removed. When the *tab method* of mounting is used (Fig. 2–27 c), the IC can be removed simply by clipping the leads just above the weld point. However, the replacement process is not as simple. You must still solder the IC lead to the tab.

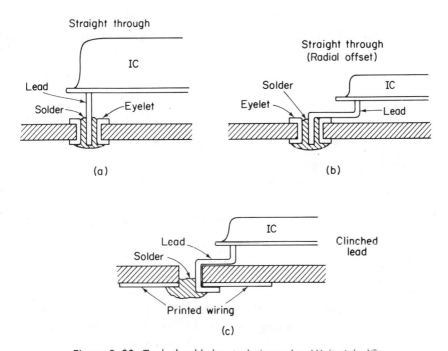

Figure 2–28. Typical soldering techniques for TO-5 style ICs.

TO-5 style packs. Figure 2–28 shows three methods for making solder connections to TO-5 style packs. From a repair standpoint the main concern is that the clinched-lead method (Fig. 2–28 c) does not require the hole to be filled with solder as do the straight-through methods. Also, you do not have to hold the IC package when soldering the leads in place. Typical mounting patterns for TO-5 style packs are shown in Fig. 2–29.

Dual in-line packs (DIP). Figure 2–30 shows a typical mounting arrangement for a DIP. Note that the solder connection methods and mounting

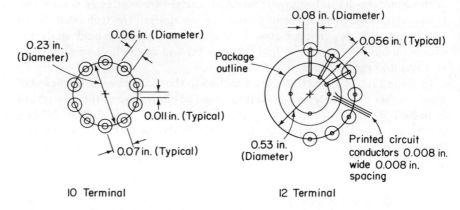

10 Terminal   12 Terminal

Figure 2–29. Typical mounting patterns for TO-5 style ICs.

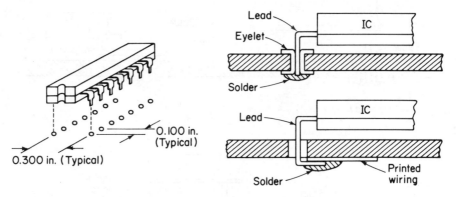

Figure 2–30. Typical mounting pattern and soldering techniques for dual in-line ICs.

patterns for DIPs are similar to those of the flat packs. However, the in-line packs are longer, permitting larger leads and more spacing between leads.

Another significant feature of the terminals for the DIP is the sharp step increase in width near the package end. This step forms a shoulder upon which the package rests when mounted on the board. Therefore, the IC package is not mounted flush against the board. As a result, it is possible to run printed circuit wiring directly under the package. From a repair standpoint it is generally easier to remove and replace in-line packages.

Lead bending. In any method of mounting ICs, you may find it necessary to bend the leads (to match the original IC that has been replaced). It is very important that the leads be supported and clamped *between the bend and the seal*. Otherwise, the seal may be broken or the lead plating may be damaged. Long-nose pliers can be used to hold the lead as shown in Fig. 2–31.

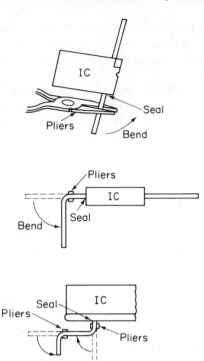

Figure 2–31. Lead-bending
techniques for ICs.

In no case should the radius of the bend be less than the diameter of the lead or, in the case of rectangular leads, less than the lead thickness. It is also important that the ends of the bent leads be perfectly straight and parallel to assure proper insertion through the holes in the printed circuit board.

2–11. SOLID-STATE SERVICING NOTES

The following notes summarize practical suggestions for servicing (troubleshooting and repair) solid-state equipment. The notes cover a wide variety of equipment.

Be sure the power to the equipment is turned off or the line cord is removed from the a-c receptacle when making in-circuit tests or repairs. Transistors can be damaged from the transients developed when changing components or inserting new transistors (in addition to the possibility of shock or short circuit). In some equipment (solid-state TV, for example) certain circuits may be *live,* even with the power switch set to *off.* To be on the safe side, pull the power plug.

When working on solid-state equipment, do not operate with any parts, such as loudspeakers or picture-tube yokes, disconnected. If the load is removed from some transistor circuits, heavy current will be drawn, resulting in possible damage to the transistor or other part such as an audio transformer.

Avoid sparks or arcs when servicing solid-state equipment. The transients developed can damage some small signal transistors. For example, when servicing solid-state TV, use a meter and high-voltage probe to measure the second anode potential. Do not arc the second anode lead to the chassis for a spark test, as is often done in vacuum-tube TV sets. Such an arc in a solid-state TV could destroy the high-voltage rectifier and possibly damage the horizontal output transistor.

If you run into an intermittent condition and can find no fault by routine checks, try tapping (not pounding) the solid-state components (transistors, diodes, etc.). If this does not produce any clues, try rapid heating and cooling of components. A small, portable hair dryer and a spray-type circuit cooler make good heating and cooling sources, respectively. First apply heat, then cool the component. The quick change in temperature will normally cause an intermittently defective component to go bad permanently. In many cases, the component will open or short, making it easy to locate. As an alternative procedure, measure the gain of a transistor with an in-circuit transistor tester. Then subject the transistor to rapid changes in temperature. If the suspected transistor changes its gain drastically or there is no gain change whatsoever, the transistor is probably defective.

If time permits, an intermittently defective transistor can be located by measuring in-circuit gain when the equipment is cold. Then let the equipment operate until the trouble occurs and measure the gain of the transistors while they are hot. Some variations will be noted in all transistors, but a leaky transistor will have a much lower gain reading when it has heated up.

If any transistor element appears to have a short (particularly the base), check the settings of any operating or adjustment controls associated with the circuit. For example, in an audio circuit, a gain or volume control set to zero or minimum can give the same indication as a short from base to ground.

Avoid checking capacitors in circuit by shunting a good capacitor across the suspected capacitor. This is especially true with an electrolytic capacitor (often used as an emitter bypass). The transient voltage surges can damage transistors. In general, avoid any short-circuit tests with solid-state equipment.

If you must service any particular make or model of equipment regularly, record the transistor gain readings of a good working unit on the schematic for future reference. Compare these gain readings against the minimum values listed in the data sheets.

Most metal case transistors have their case tied to the collector. Therefore, you can use the case as a test point. Avoid using a clip-type probe on transistors. Also avoid clipping onto some of the subminiature resistors used in solid-state equipment. Any subminiature components can break with rough handling.

When injecting a signal into the base of a transistor, make sure there is a blocking capacitor in the signal generator output. Most signal generators have some form of blocking capacitor to isolate the output circuit from the d-c voltages that may appear in the circuit. In the case of solid-state equipment, the blocking capacitor also prevents the base from being returned to ground (through the generator output circuit) or from being connected to a large d-c voltage (in the generator circuit). Either of these conditions can destroy the transistor. If the generator is not provided with a built-in blocking capacitor, connect a capacitor between the generator output lead and the transistor base.

2–12. SOLID-STATE OSCILLATOR BIAS PROBLEMS

One of the problems in troubleshooting solid-state oscillator circuits is the variation in bias arrangements. There are three basic bias schemes. Most AF oscillators are biased class A, where the transistor remains forward-biased at all times. RF oscillators are generally reverse-biased so that they conduct on half-cycles. With blocking oscillators, the transistor is reverse-biased for most of the time (sometimes 90 per cent of the time) but conducts heavily once each cycle. The obvious problem here is how does the transistor conduct initially in a free-running oscillator?

In most free-running, solid-state oscillator circuits, the transistor is initially forward-biased by d-c voltages (through the bias-resistance network). This turns the transistor on so that the collector circuit starts to conduct. Feedback occurs and the transistor is driven into heavy conduction. At the same time, a capacitor connected to the base is charged in the forward-bias condition. When saturation is reached, there is no further feedback and the capacitor discharges. This reverse-biases the transistor and maintains the reverse bias until the capacitor has discharged to a point where the fixed forward bias again causes conduction. This feature presents a problem in design of class C, solid-state RF oscillators. If the capacitor is too large, it may not discharge in time for the next half-cycle. In that case, the class C oscillator acts as a blocking oscillator, where the frequency is controlled by the capacitance and resistance of the circuit. If the capacitor is too small, the class C oscillator may not start at all.

From a practical troubleshooting standpoint, the measured condition of bias on a solid-state oscillator can provide a good clue as to operation, provided you know how the oscillator is to operate.

Figures 2–32, 2–33, and 2–34 are schematic diagrams of solid-state, audio,

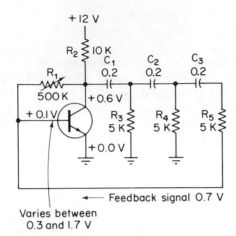

Feedback signal 0.7 V

Varies between
0.3 and 1.7 V

Figure 2–32. Class A audio oscillator (forward-biased at all times).

RF, and blocking oscillators, respectively. The following paragraphs describe operation of these circuits, in relation to troubleshooting analysis.

The oscillator of Fig. 2–32 is forward-biased at all times through R_1 and is typical of low-frequency circuits where a good sinewave is required. Resistor R_1 is adjusted (or selected) so that the base will remain positive at all times (forward-biased) even at the peak of the negative feedback signal. Thus, in troubleshooting a circuit similar to the one in Fig. 2–32, anything but a forward-bias condition (reverse-bias or zero-bias) is an indication of trouble.

The oscillator of Fig. 2–33 is initially forward-biased through R_1 and R_2. As Q_1 starts to conduct and in-phase feedback is applied to the emitter (to sustain oscillation), capacitor C_1 starts to charge. When saturation is reached (or approached) and the feedback stops, capacitor C_1 then discharges in the opposite polarity, reverse-biasing Q_1. Capacitor C_1 is selected (in value) so that it discharges to a voltage less than the fixed forward bias, *before* the next half-cycle. Thus, transistor Q_1, conducts on slightly less than the full half-cycle. Typically, a class C RF oscillator such as in Fig. 2–33 will conduct on about 140° of the 180° half-cycle.

When the circuit of Fig. 2–33 is oscillating normally, the *average voltage value* between emitter and base will be such that the transistor is reverse-biased or possibly zero-biased, even though the fixed bias is forward. That is, if the oscillator is on, you will measure a reverse bias or zero bias. Thus, in troubleshooting a circuit such as in Fig. 2–33, a forward-bias condition is an indication of trouble.

While on the subject of bias, it is commonly assumed that transistor junctions (and diodes) start to conduct as soon as forward voltage is applied. This is not true. Figure 2–35 shows characteristic curves for three different types of transistor junctions. All three junctions are silicon, but the same condition exists for germanium. None of the junctions conducts noticeably at

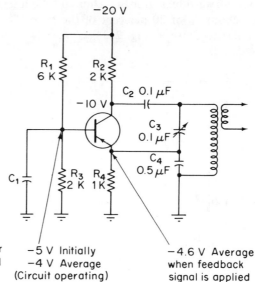

Figure 2–33. Class C RF oscillator (reverse-biased or zero-biased with circuit operating).

−5 V Initially
−4 V Average
(Circuit operating)

−4.6 V Average when feedback signal is applied

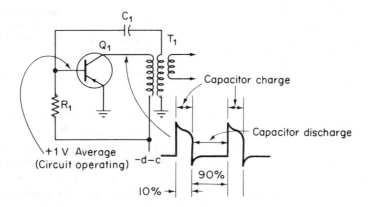

Figure 2–34. Blocking oscillator (reverse-biased with circuit operating).

0.6 V. Current starts to rise at 0.7 V. At 0.8 V, one junction draws almost 80 mA. At 1 V, the d-c resistance is on the order of 2 or 3 Ω, and the transistor draws almost 1 A. In a germanium transistor, noticeable current flow will occur at about 0.3 V.

The blocking oscillator of Fig. 2–34 is initially forward-biased by the charge of C_1 through R_1. As Q_1 starts to conduct and in-phase feedback is applied through C_1 to the base, Q_1 is driven rapidly into saturation. Capacitor C_1 charges to a high value during this period. When saturation is reached and there is no further feedback, capacitor C_1 discharges, thereby reverse-biasing

Q_1. The value of C_1 is chosen so that it discharges to a value less than the fixed forward bias over the required time interval. Typically, Q_1 will conduct for about 10 or 20 per cent of the time. This produces output waveforms similar to those in Fig. 2–34.

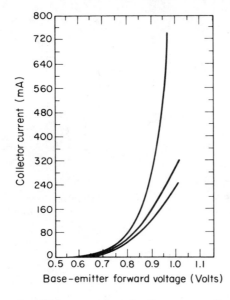

Base–emitter forward voltage (Volts)

Figure 2–35. Characteristic curves for silicon transistor junctions (collector current flow versus base-emitter forward voltage).

It will be seen that the frequency of a blocking oscillator is determined by the values of C_1 and R_1 and the value of the fixed bias.

As in the case of the RF circuit, when the blocking oscillator in Fig. 2–34 is operating normally, the average base-emitter voltage will be such that you will measure a reverse bias. Thus, in troubleshooting a circuit as shown in Fig. 2–34, a forward-bias condition (and usually a zero-bias condition) is an indication of trouble.

To sum up, bias measurements provide a clue as to the performance of solid-state oscillators. However, bias measurements do not provide proof positive. The one sure test of an oscillator is measurement of the waveform on an oscilloscope. If the waveform is present and is of the right shape, amplitude, and time duration (or frequency), the oscillator is operating properly.

Usually, the collector is the best place to measure a solid-state oscillator waveform. Keep in mind that the oscilloscope must have a bandpass greater than the oscillator frequency. If not, the waveform will be distorted or will not pass at all.

If it is necessary to measure a frequency higher than the oscilloscope bandpass, use an RF probe to test the oscillator. Such probes (available with most laboratory oscilloscopes) convert the RF voltage into a d-c voltage suitable for measurement on the oscilloscope. An RF probe can also be used with a conventional d-c voltmeter.

2–13. EFFECTS OF LEAKAGE ON AMPLIFIER GAIN

When there is considerable leakage in a solid-state amplifier, the gain will be reduced to zero and/or the signal waveform will be drastically distorted. Such a condition will also produce abnormal waveforms and transistor element voltages. These indications will make the problem easy, or relatively easy, to locate. The real difficulty occurs when there is just enough leakage to reduce amplifier gain but not enough leakage to seriously distort the waveform or produce transistor voltages that are way off.

Collector-base leakage is the most common form of transistor leakage and produces a classic condition of low gain. When there is any collector-base leakage, the transistor will be forward-biased, or the forward bias will increase. This condition is shown in Fig. 2–36. Collector-base leakage has the

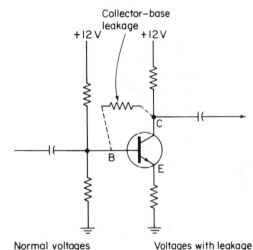

Normal voltages	Voltages with leakage
C – 6 V	C = 4 V
E = 2 V	E = 3 V
B = 2.5 V	B = 3.5 V

Figure 2–36. Effect of collector-base leakage on transistor element voltages.

same effect as connecting a resistance between the collector and base. The base assumes the same polarity as the collector (although at a lower value) and the transistor is forward-biased. If leakage is sufficient, the forward bias will be enough to drive the transistor into or near saturation. When a transistor is operated at or near the saturation point, the gain is reduced. This is shown in the curve of Fig. 2–37.

If the normal transistor element voltages are known (from the service

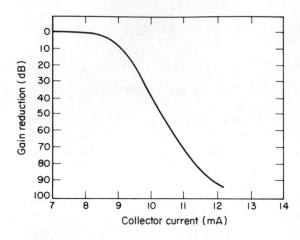

Figure 2–37. Relative gain of solid-state amplifier at various average collector current levels.

literature), excessive transistor leakage can be spotted easily since all of the transistor voltages will be off. For example in Fig. 2–36, the base and emitter will be high and the collector will be low when measured in reference to ground. However, if the normal operating voltages are not known, the transistor can appear to be good since all of the *voltage relationships* are normal. That is, the collector-base junction is reverse-biased (collector more positive than base for an NPN) and the emitter-base junction is forward-biased (emitter less positive than base for NPN).

A simple way to check transistor leakage is shown in Fig. 2–38. Measure the collector voltage to ground. Then short the base to the emitter and remeasure the collector voltage. If the transistor is not leaking, the base-emitter short will turn the transistor off and the collector voltage will rise to

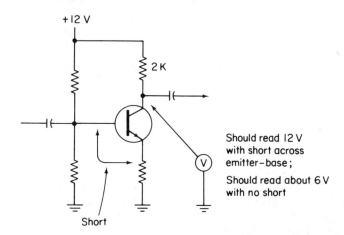

Figure 2–38. Checking for transistor leakage in circuit.

the same value as the supply. If there is any leakage, a current path will remain through the emitter resistor, emitter-base short, collector-base leakage path, and collector resistor. There will be some voltage drop across the collector resistor, and the collector will have a voltage some value lower than the supply.

Note that most meters will draw some current, and this current will pass through the collector resistor. This can lead to some confusion, particularly if the meter draws heavy current (has a low ohms-per-volt rating). To eliminate any doubt, connect the meter to the supply through a resistor with the same value as the collector resistor. The drop, if any, should be the same as when the transistor collector is measured to ground. If the drop is much lower when the collector is measured, the transistor is leaking.

As an example, assume that in the circuit of Fig. 2–38 the supply is 12 V, the collector resistance is 2 K, and the collector measures 4 V with respect to ground. This means that there is an 8-V drop across the collector resistor and a collector current of 4 mA $(8/2000 = 4 \text{ mA})$. Normally, the collector is operated at about one-half of the supply voltage, or 6 V. However, simply because the collector is at 4 V instead of 6 V does not make the circuit faulty. Some circuits are designed that way. Therefore, the transistor must be checked for leakage. Now assume that the collector voltage rises to 10 V when the base and emitter are shorted. This indicates that the transistor is cutting off, but there is still some current flow through the collector resistor, about 1 mA $(2/2000 = 1 \text{ mA})$.

A 1-mA current flow is high for a meter. However, to confirm a leaking transistor, connect the meter through a 2-K resistor to the 12-V supply, preferably at the same point where the collector resistor connects to the supply. Now assume that the indication is 11.7 V through the external resistor. This indicates that there is some transistor leakage. The amount can be estimated as follows: $11.7 - 10.5 = 1.2$-V drop; $1.2/2000 = 0.6$ mA. However, from a practical troubleshooting standpoint, the presence of any current flow with the transistor supposedly cut off is sufficient cause to replace the transistor.

2–14. CAPACITORS IN SOLID-STATE CIRCUITS

Although the functions of capacitors in solid-state circuits are similar to those in vacuum-tube equipment, the results produced by capacitor failure are not necessarily the same. An emitter bypass capacitor is a good example. The emitter resistor in a solid-state circuit (such as R_4 in Fig. 2–39) is used to stabilize the transistor d-c gain and prevent thermal runaway. With an emitter resistor in the circuit, any increase in collector current produces a greater drop in voltage across the resistor. When all other factors remain the

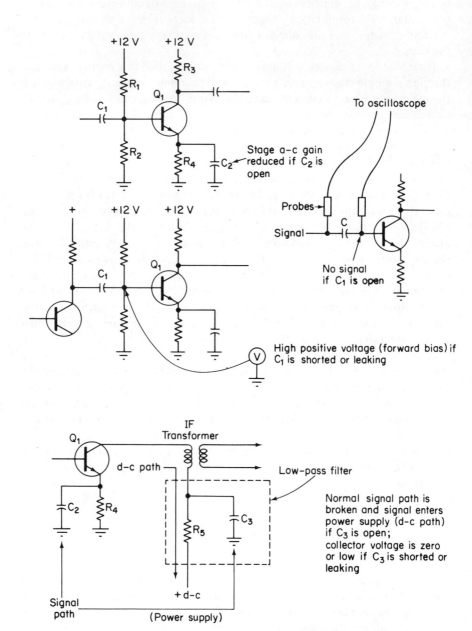

Figure 2–39. Effects of capacitor failure in solid-state circuits.

same, this change in emitter voltage reduces the base-emitter forward-bias differential, thus tending to reduce collector current flow.

When circuit stability is more important than gain, the emitter resistor is not bypassed. When a-c or signal gain must be high, the emitter resistance is bypassed to permit passage of the signal. If the emitter bypass capacitor is open, stage gain is reduced drastically, although the transistor d-c voltages remain substantially the same. Thus, if there is a low-gain symptom in any solid-state amplifier with emitter bypass and the voltages appear normal, check the bypass capacitor. This can be done by shunting the bypass with a known good capacitor of the same value. As a precaution, shut off the power before connecting the shunt capacitor; then reapply power. This will prevent damage to the transistor (due to large current surges).

The functions of coupling and decoupling capacitors in solid-state circuits are essentially the same as for vacuum-tube equipment. However, the capacitance values are much larger, particularly at low frequencies. Electrolytics are usually required to get the large capacitance values. From a practical standpoint, electrolytics tend to have more leakage than mica or ceramic capacitors. However, good-quality electrolytics (typically the bantam type found in solid-state) will have leakage of less than 10 μA at normal operating voltage.

The function of C_1 in Fig. 2–39 is to pass signals from the previous stage to the base of Q_1. If C_1 is shorted or leaking badly, the voltage from the previous stage is applied to the base of Q_1. This forward-biases Q_1, causing heavy current flow and possible burnout of the transistor. In any event, Q_1 is driven into saturation, and stage gain is reduced. If C_1 is open, there will be little or no change in the voltages at Q_1, but the signal from the previous stage will not appear at the base of Q_1. From a troubleshooting standpoint, a shorted or leaking C_1 will show up as abnormal voltages (and probably as distortion of the signal waveform). If C_1 is suspected of being shorted or leaky, replace C_1. An open C_1 will show up as a lack of signal at the base of Q_1, with a normal signal at the previous stage. If an open C_1, is suspected, replace C_1 or try shunting C_1 with a known good capacitor, whichever is convenient.

The function of C_3 in Fig. 2–39 is to pass operating signal frequencies to ground (to provide a return path) and to prevent signals from entering the power supply line or other circuits connected to the line. In effect, C_3 and R_5 form a low-pass filter that passes d-c and very-low-frequency signals (well below the operating frequency of the circuit) through the power supply line. Higher-frequency signals are passed to ground and do not enter the power supply line.

If C_3 is shorted or leaking badly, the power supply voltage will be shorted to ground or greatly reduced. This reduction of collector voltage will make the stage totally inoperative or will reduce the output, depending on the amount of leakage in C_3.

If C_3 is open, there will be little or no change in the voltages at Q_1. However, the signals will appear in the power supply line. Also, signal gain will be reduced and the signal waveform will be distorted. In some cases, at higher signal frequencies, the signal simply cannot pass through the power supply circuits. Since there is no path through an open C_3, the signal will not appear on the collector circuit in any form. From a practical standpoint, the results of an open C_3 will depend on the values of R_5 (and other power supply components) as well as on the signal frequency involved.

2-15. EFFECTS OF LOW VOLTAGE ON RESISTANCE AND COLD SOLDER JOINTS

The effects of shorts on resistors in solid-state circuits are less drastic than in vacuum-tube circuits. This is because of the lower voltages used in solid-state. For example, most solid-state circuits operate at voltages well below 25 V, typically 12 V or less. A 1-k load or bias network resistance (which is typically a very low value) shorted directly across a 25-V source (a very high value) produces only 25 mA current flow, or about 0.6 W. A 1-W resistor can easily handle this power with no trouble. Even a 0.5-W resistor would probably survive a temporary short of this level. For that reason, resistors do not burn out as often in solid-state equipment, nor do resistance values change due to prolonged heating. There are exceptions of course, but most solid-state troubles are the result of defects in capacitors (first), transistors (second), and diodes (third).

The low voltages in solid-state equipment have just the opposite effect on cold solder joints and partial breaks in printed wiring. Often the high voltages in vacuum-tube equipment can overcome the resistance created by cold solder joints and partial printed circuit breaks. When there is no obvious cause for a low voltage at some point in the circuit or there is an abnormally high resistance, look for cold solder joints or defects in printed circuit wiring.

Use a magnifying glass to locate defects in printed circuit wiring. Sometimes minor breaks in printed wiring can be repaired by applying solder at the break, observing all precautions described in Sec. 2–10. 1. However, this is recommended only as a temporary measure. Under emergency conditions, it is possible to run a wire between two points on either side of the break. However, it is recommended that the entire board be replaced as soon as it is practical.

Cold solder joints can sometimes be found with an ohmmeter. Remove all power. Connect the ohmmeter across two wires leading out of the suspected cold solder joint as shown in Fig. 2–40. Flex the wires by applying pressure

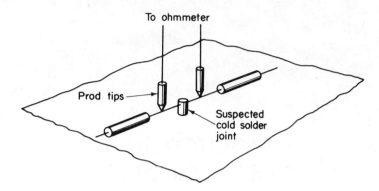

Figure 2–40. Locating cold solder joints with an ohmmeter.

with the ohmmeter prod tips. Switch the ohmmeter to different ranges and check if there is any change in resistance. For example, a cold solder joint can appear to be good on the high ohmmeter ranges but as an open on the lower ranges. Look for resistance indications that tend to drift or change when the ohmmeter is returned to a particular scale. If a cold solder joint is suspected, reheat the joint with a soldering tool; then recheck the resistance.

3. INTRODUCTION TO DIGITAL CIRCUITS

Obviously, it is necessary to have a knowledge of digital circuits before attempting to service any type of digital equipment. The data in this section are provided as a summary or refresher but not as a basic course in digital circuits or logic symbology. However, if the reader understands the data in this section, he should have no difficulty in working with the logic diagrams and schematics found in the instruction manuals of digital equipment manufacturers.

3–1. LOGIC SYMBOLOGY

Logic symbols and diagrams are a form of language used in digital circuits. One major problem found in servicing digital equipment is that each manufacturer speaks a different dialect of the same language. No two major manufacturers use identical symbols in the instruction manuals to represent the same circuit element.

This problem was supposed to have been eliminated when MIL-STD-806 was introduced. However, many manufacturers still do not follow this government standard exactly, some with good reason. For example, the logic symbols used by Hewlett-Packard in their instruction manuals clarify many points not discussed by MIL-STD-806. Throughout the following sections, the military standard symbols will be used (where a MIL-STD-806 symbol exists). Manufacturers' modifications will also be given as an alternative where they clarify operation of the circuit or symbol function.

98

3–2. BINARY LOGIC

In the binary system (common to most digital equipment encountered by service technicians), information is broken down into elementary *bits*. For example, all numbers can be made up using only *zeros* and *ones* rather than zero through nine as in the familiar decimal system. Consequently, instead of requiring ten different values to represent one digit, digital equipment using the binary method needs only *two values* (thus the term binary) for each digit.

Each bit in the binary system can exist in only one of two possible states. For logic, these states are designated "1" and "0". A 1 means *yes, assertion, enable,* or *true.* Conversely, 0 means *no, negation, disable inhibit,* or *false.* Use of the words *true* and *false* does not imply that one state is more important than the other. The states are conditions, and both states are equally significant and are used equally in a two-state system.

A single bit is always used to represent a function in binary logic. For example, suppose the information to be conveyed is the presence (or absence) of a count. If the count has been received (or perhaps stored), the bit representing the count will be in the 1 state for presence or truth; otherwise the bit will be in the 0 state, denoting absence or falsity.

The word bit is used to denote

1. An assertion or negation of a variable.
2. Storage of a variable.

In electronic equipment, true and false states are generally represented by voltage or current levels and must be defined. For example, true could be $+12$ V, false could be -12 V; true could be the presence of current, false could be the absence of current. Generally, voltage levels rather than current levels are used to define the true and false states.

In some (but not all) logic diagrams, a $(+)$ or $(-)$ sign may be used within any logic symbol to define the true state for that element in greater detail. A $(+)$ sign within a logic symbol means the *relatively positive level* of the two logic voltages at which that circuit operates and is said to be true. Note that the true voltage level does not have to be absolutely positive (that is, above ground or above 0-V reference).

The two voltage levels at which a logic circuit operates could be -10 V and -5 V. A $(+)$ sign within the logic symbol representing such a circuit will indicate that the -5 V level is true and the -10-V level is false. This is because -5 V is closer to positive than -10 V. A $(-)$ sign within the symbol will indicate that the -10-V level is true, since it is more negative, and that the -5 V level is therefore false.

The sign should be used for all logic elements in which true and false levels are meaningful. As an alternative, the logic diagrams can state in a note that the logic is either all positive-true or all negative-true. (MIL-STD-806 does not distinguish positive-true from negative-true.)

3-3. BINARY NUMBERS

In digital equipment, numbers are usually represented in binary form. The binary bits representing the number are assigned a *weight* or value; thus the number of bits required depends on the magnitude of the number to be represented. Because each bit can exist in only two states (1 and 0), the weights assigned to each successive bit can increase by a maximum factor of 2. Thus, in a pure binary-form number, the weights assigned to successive bits are

32	16	8	4	2	1	.	1/2	1/4	1/8	1/16
2^5	2^4	2^3	2^2	2^1	2^0	.	2^{-1}	2^{-2}	2^{-3}	2^{-4}

Since each bit is either true or false, the *weights of all true bits are added* to obtain the number. For example, the decimal number 22.5 in pure binary is 10110.1000.

Binary weight	2^4	2^3	2^2	2^1	2^0	.	2^{-1}	2^{-2}	2^{-3}	2^{-4}
Decimal weight	16	8	4	2	1	.	1/2	1/4	1/8	1/16
Binary weight	1	0	1	1	0	.	1	0	0	0

By adding all of the true bits $(16 + 4 + 2 + \frac{1}{2})$ the total number (22.5) may be found.

If all of the possible bits are true (given a binary number 11111.1111), the decimal number represented is $16 + 8 + 4 + 2 + \frac{1}{2} + \frac{1}{4} + \frac{1}{8} + \frac{1}{16}$, or 31 and $\frac{15}{16}$. If the number 32 is needed, the 2^5-bit must be added to the left of the decimal point. (In the pure binary system, the decimal point is called the *binary point*.) Note that bits can also be added to the right of the binary point, thus increasing resolution.

3-4. BINARY-CODED DECIMAL SYSTEM

The binary-coded decimal (BCD) system combines the advantages of the binary system (the need for only two states, 1 or 0) and the convenience of the familiar decimal representation. In the BCD system, a number is expressed in normal decimal coding, but each digit in the number is expressed in binary form.

For example, the number 37 in pure binary BCD form would appear as follows.

BCD	Tens Digit	Units Digit
(Pure binary weight)	0011 (8421)	0111 (8421)
Decimal	3	7

Four bits are needed for each digit. In general, four bits yield 2^4 for 16 possible combinations, as shown in Table 3–1.

TABLE 3–1. TYPICAL BCD CODES USED IN DIGITAL EQUIPMENT

Code Bits	Decimal Equivalents			
	8421	42*21	XS-3	2421
0000	0	0		0
0001	1	1		1
0010	2	2		2
0011	3	3	0	3
0100	4		1	4
0101	5		2	5
0110	6	4	3	6
0111	7	5	4	7
1000	8		5	
1001	9		6	
1010	10		7	
1011	11		8	
1100	12	6	9	
1101	13	7		
1110	14	8		8
1111	15	9		9

Note that there are many BCD codes used other than the *pure binary* 8421 code. Three typical codes (42*21, XS-3, and 2421) are shown in the table. In every case, *four bits* are required for each digit. Likewise, six combinations of four bits are unused in each code. The six unused combinations are often referred to as *forbidden codes*.

Besides the pure binary and other BCD forms, it is possible to express numbers in digital equipment applications by other means. One such system is the *10-line code,* where each digit is represented by ten bits, with each bit weighted 0 through 9. For a given digit, only one bit of the 10-line code can be true at one time. This 10-line code is sometimes known as multiple-line code. A variation of this system is used in electronic counter readouts, as described in Sec. 3–20.

The *negation* or *not* function is also used in BCD systems. The not condition is indicated by a bar above the bit identification. For example, a *not eight* is $\bar{8}$. For all numbers where the 8-bit is true, the $\bar{8}$-bit is by definition false. Note that the $\bar{8}$-bit is a separate bit but is related by definition to the 8-bit. Conversely, for all numbers where the 8-bit is false (not true), the $\bar{8}$-bit is true.

3–5. BASIC DIGITAL LOGIC ELEMENTS AND SYMBOLS

Although there are many variations of digital logic elements and their symbols, there are only four basic classes or groups. These are *gates, amplifiers, switching elements,* and *delay elements.*

2–5.1. Gates

A gate is a circuit which produces an output on condition of certain rules governing input combinations. (The specific rules for various gates are discussed in Sec. 3–7 through 3–12.)

As shown in Fig. 3–1, the basic gate symbol has input lines connecting to the flat side of the symbol and output lines connecting to the curved side. Since inputs and outputs are easily identifiable in this manner, the symbol can be shown facing left or facing right (or facing up or down) as necessary.

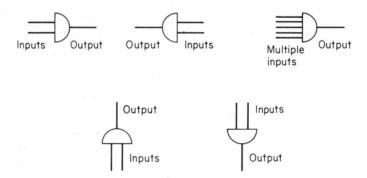

Figure 3–1. Basic gate symbols.

3–5.2. Amplifiers

Amplifiers are not necessarily limited to digital work. However, when so used, the driving or input signals will normally be digital (pulsed, binary, etc.). Consequently, the output of the amplifier will be an amplified and otherwise modified form of the input.

As shown in Fig. 3–2, the amplifier symbol is an equilateral triangle, with the input applied to the center of one side and the output connected to the opposite point of the triangle. As with gates, the amplifier may be shown in any of the four positions.

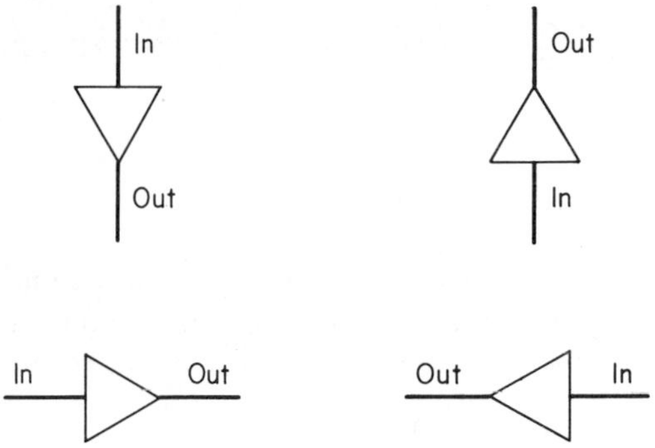

In

Out

Out

In

In   Out

Out   In

Figure 3–2. Basic amplifier symbols.

3–5.3. Switching Elements

Switching elements used in digital work are a form of the *multivibrator:* bistable (flip-flop, Schmitt trigger), monostable (one-shot), and astable (free-running multivibrator). A description of a typical solid-state multivibrator is given in Sec. 3–15.

According to the type of switching circuit, inputs cause the state of the circuit to switch, reversing the outputs. That is, an output formerly true will switch to false, and vice versa.

As shown in Fig. 3–3, the most common basic symbol for switching circuits is a horizontal rectangle divided horizontally, with the upper portion representing the *set side* and the lower portion representing the *reset side*. A switching element is said to be *set* when the output from the set side is true. The element is *reset* when the output from the reset side is true.

Figure 3–3. Basic switching-element symbol.

Inputs are on the left; outputs are on the right. To avoid confusion, switching elements are always drawn facing the same way (or should be so drawn).

3–5.4. Delay Elements

A delay element provides a *time delay* between input and output signals. As shown in Fig. 3–4, the symbol accepts the input on the left and provides output on the right. Like switching elements, delay elements are always drawn facing the same way.

Input ————⸦⸧———— Output   Figure 3–4. Basic delay-element symbol.

3–6. MODIFICATION OF LOGIC SYMBOLS

Basic logic symbols are usually modified to express circuit conditions. Although each manufacturer may have its own set of modifiers together with those of MIL-STD-806, the following modifiers are in general use.

3–6.1. Truth Polarity

As discussed previously, positive (+) or negative (−) indicators may be placed inside a symbol to designate whether the true state for that

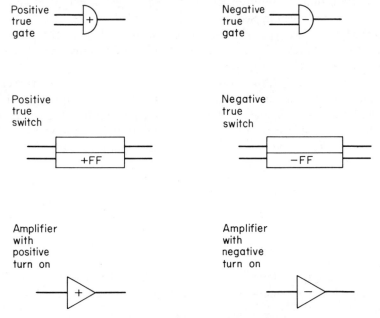

Figure 3–5. Truth polarity in symbols.

circuit is positive or negative *relative to* the false state. This is done frequently with gates and switching elements, as shown in Fig. 3–5. (The FF designation in Fig. 3–5 indicates that the particular switching element is a flip-flop.)

Where all symbols on a particular diagram have the same polarity, a note to the effect that all logic is positive true or all negative true may be used instead of having individual polarity signs in each symbol. Polarity signs used in amplifier symbols do not have any direct logic significance. Rather, the polarity signs are a troubleshooting aid, indicating the *polarity required to turn the amplifier on.*

As shown in Fig. 3–5, the positive-true gate and the positive FF operate with true levels being positive with respect to false levels. Similarly, the negative-true gate and the FF operate with true levels being negative with respect to the false levels.

3–6.2. Inversion

Generally, logic inversion is indicated by an *inversion dot* at inputs or outputs. (In some cases, *inverted pulses* are shown at the inputs and outputs.) When the inversion dot appears on an input (generally only on gates and switching elements), the input will be effective when the input signal is of opposite polarity to that *normally required.* For example, if the switching element in Fig. 3–6 is normally positive true, or is used on a diagram where all logic is positive true, a negative input at the inversion dot will set the circuit.

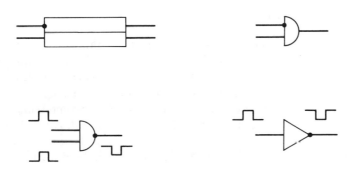

Dots may be solid or open

Figure 3–6. Methods of indicating inversion in symbols.

When the inversion dot appears at an output, generally only on gates and amplifiers, the output will be of opposite polarity to that *normally delivered.* For example, if the gate of Fig. 3–6 is used in a positive-true logic circuit, the output will be negative. Likewise, the amplifier in Fig. 3–6 will produce a positive output if the input is negative, and vice versa.

3–6.3. A-C Coupling

Capacitor inputs to logic elements are indicated by an arrow as shown in Fig. 3–7. In the case of gates and switching elements, the element responds only to a change of the a-c coupled input in the true-going direction. An inversion dot used in conjunction with the coupling arrow indicates that the element responds to a change in the *false-going* direction.

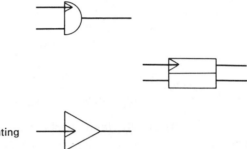

Figure 3–7. Methods of indicating a-c coupling in symbols.

In the case of an amplifier, a pulse edge of the same polarity as given in the symbol turns the amplifier on briefly, and then off as the capacitor discharges. The output is then a pulse of the same width as the amplifier on time. With an inversion dot at the amplifier output, the output pulse is inverted.

3–7. AND GATES

The symbol, basic circuit, and truth table for the AND gate are shown in Fig. 3–8. By definition, for output C to be true, both inputs A *and* B must be true, hence the term AND gate. This can be illustrated by means of the truth table in Fig. 3–8. (Similar truth tables are found in many logic diagrams for digital equipment.) In this table, 1 represents true and 0 represents false. Note that C is true (1) only when both A and B are true, and that the truth table does not define true and false. That is, the table does not spell out positive true, negative true, the voltage level for true or false, and the like.

If the AND gate is positive true, a (+) sign should be given in the symbol. A negative-true AND gate should have a (−) sign in the symbol. (Note that MIL-STD-806 does not distinguish positive true from negative true.)

Assume that the respective true/false levels for the two gates in Fig. 3–8 are +5 V/0 V and −5 V/0 V, as shown. These two values can be substituted for 1 and 0 as in Table 3–2.

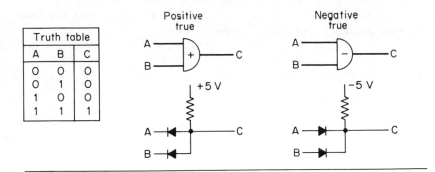

Truth table		
A	B	C
0	0	0
0	1	0
1	0	0
1	1	1

Positive true

Negative true

+5 V

−5 V

Multiple input

Truth table			
A	B	C	D
0	0	0	0
0	0	1	0
0	1	0	0
0	1	1	0
1	0	0	0
1	1	0	0
1	1	1	1

MIL−STD−806

Basic

A and B high = F high

Mixed

A low and B high = F high

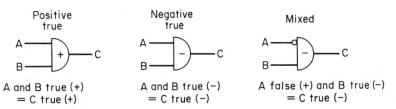

Hewlett−Packard

Positive true

A and B true (+)
= C true (+)

Negative true

A and B true (−)
= C true (−)

Mixed

A false (+) and B true (−)
= C true (−)

Figure 3–8. AND gate symbols.

In both cases, the true voltage level appears at output C when the A and B input diodes are reverse-biased by true inputs. At all other times, the false level (approximately 0 V) exists at output C.

TABLE 3–2. AND GATE TRUTH TABLES

Positive Gate				Negative Gate		
A	B	C		A	B	C
0 V	0 V	0 V		0 V	0 V	0 V
0 V	+5 V	0 V		0 V	−5 V	0 V
+5 V	0 V	0 V		−5 V	0 V	0 V
+5 V	+5 V	+5 V		−5 V	−5 V	−5 V

For example, in the positive-true gate, if A is at 0 V, current will flow through the corresponding diode, lowering the voltage at C to approximately 0 V. The same event will occur if B (or both A and B) is at 0 V. If both A and B are at +5 V, no current will flow through either diode, and C will rise to approximately +5 V.

AND gates are not restricted to two inputs, but may have any number of inputs, including one input in special cases. (Single-input AND gates require that the input be of a given polarity and/or level to produce a true output. Single-input gates are not truly AND gates but are often so described in the instruction manuals of digital equipment manufacturer's.)

For multiple-input AND gates, the rule is that *all inputs must be true for the output to be true,* as shown in Fig. 3–8.

3–8. OR GATES

The symbol, basic circuit, and truth table for the OR gate are shown in Fig. 3–9. By definition, for C to be true, either input A *or* input B must be true, hence the term OR gate. This is illustrated by the truth table of Fig. 3–9. Note that C is true (1) whenever any of the inputs are true and that the truth table does not define true and false. If a (+) sign or a (−) sign is given in the symbol, the OR gate is defined as positive true or negative true, respectively.

Note that the circuit for a negative-true OR gate is the same as for a positive-true AND gate, and vice versa. This produces inversion of the two circuit functions. For example, the 0-V level is now defined as true, while the +5-V and −5-V levels are false for the two gates in Fig. 3–9. These two values can be substituted for 1 and 0 as in Table 3–3.

Truth table		
A	B	C
0	0	0
0	1	1
1	0	1
1	1	1

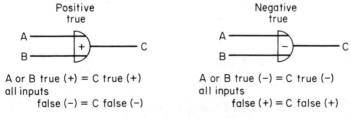

Multiple input

Truth table			
A	B	C	D
0	0	0	0
0	0	1	1
0	1	0	1
0	1	1	1
1	0	0	1
1	0	1	1
1	1	0	1
1	1	1	1

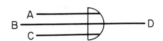

MIL–STD–806

A or B high = F high   A or B low = F low

Hewlett – Packard

Positive true   Negative true

A or B true (+) = C true (+)
all inputs
      false (−) = C false (−)

A or B true (−) = C true (−)
all inputs
      false (+) = C false (+)

Figure 3–9. OR gate symbols.

TABLE 3–3. OR GATE TRUTH TABLES

Positive Gate				Negative Gate		
A	B	C		A	B	C
−5 V	−5 V	−5 V		+5 V	+5 V	+5 V
−5 V	0 V	0 V		+5 V	0 V	0 V
0 V	−5 V	0 V		0 V	+5 V	0 V
0 V	0 V	0 V		0 V	0 V	0 V

As with AND gates, OR gates may have more than two inputs. For multiple-input OR gates, the rule is that *any true input will produce a true output,* as shown in Fig. 3–9.

3–9. NAND GATES

The symbol, basic circuit, and truth table for the NAND gate are shown in Fig. 3–10. The NAND gate is a variation of the conventional

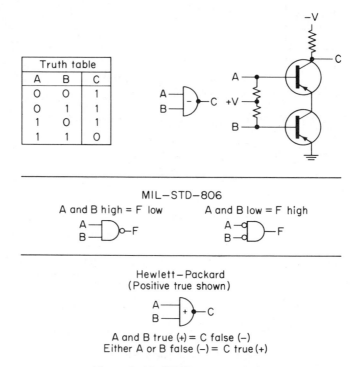

Figure 3–10. NAND gate symbols.

AND gate, delivering an *inverted* (false) output when *all inputs are true*. When either or both inputs are false, the output is true. The term NAND is a contraction of *not and*. A NAND gate uses an active inverting element in the gate circuitry and may have any number of inputs.

Operation of the NAND gate in Fig. 3–10 is as follows. For output C to be positive (at or near ground level), both Q_1 and Q_2 must be conducting. This requires that *both* A and B inputs be negative. If either A or B is positive, then either Q_1 or Q_2 is biased off, and the output is negative (C will assume the level of $-V$).

3–10. NOR GATES

The symbol, basic circuit, and truth table for the NOR gate are shown in Fig. 3–11. A NOR gate is a variation of the conventional OR gate, delivering an *inverted* (false) output when *any or all of its inputs are true*. When all the inputs are false, the output is true. The term NOR is a contraction of *not or*. A NOR gate uses an active inverting element in the gate circuitry and may have any number of inputs.

Operation of the NOR gate in Fig. 3–11 is as follows. For output C to be

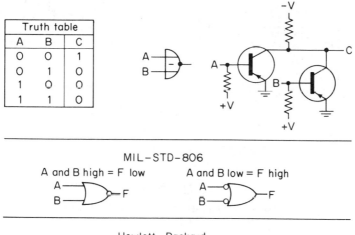

Truth table		
A	B	C
0	0	1
0	1	0
1	0	0
1	1	0

MIL–STD–806

A and B high = F low   A and B low = F high

Hewlett–Packard
(Positive true shown)

Either A or B true (+) = C false (−)
All inputs false (−) = C true (+)

Figure 3–11. NOR gate symbols.

positive (at or near ground level), either or both Q_1 and Q_2 must be conducting. This requires that either or both inputs A and B are negative. If both A and B are positive, Q_1 and Q_2 are biased off, and the output is negative (C will assume the level of $-V$).

3–11. EXCLUSIVE OR GATES

The symbol, basic circuit, and truth table for the EXCLUSIVE OR gate are shown in Fig. 3–12. An EXCLUSIVE OR gate is a special type of OR gate. It has two inputs: the output will be true if one *but not both* of the inputs is true. The converse statement is equally accurate: the output will be false if the inputs are *both true* or *both false*. The EXCLUSIVE OR gate is independent of polarity and generally is not spoken of as being either positive true or negative true.

Operation of the EXCLUSIVE OR gate in Fig. 3–12 is as follows. If either A or B is positive, Q_1 is biased on via CR_1 or CR_2, and the C output is positive. If A and B are both negative, Q_1 is biased on, but the voltage at

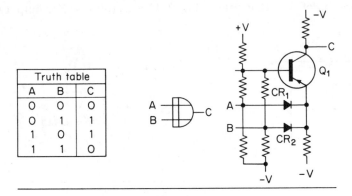

Truth table		
A	B	C
0	0	0
0	1	1
1	0	1
1	1	0

MIL–STD–806

F is high if (and only if) any one input is high, and all other inputs are low.

Hewlett–Packard

One input (A or B) true makes C true. Both inputs (A and B) true or false makes C false

Figure 3–12. EXCLUSIVE OR gate symbols.

the Q_1 emitter is negative and the C output is negative. With A and B both positive, Q_1 is biased off and the C output remains negative. (These results require proper selection of resistor values and operating voltages.)

3–12. ENCODE GATES

The symbol, basic circuit, and truth table for the ENCODE gate are shown in Fig. 3–13. The ENCODE gate has one input and multiple outputs. When the input is true, all outputs are true. When the input is false, the output may be true or false, depending on the state of the logic element to which they are connected.

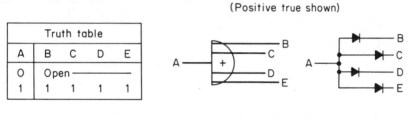

(Positive true shown)

Truth table				
A	B	C	D	E
0	Open			
1	1	1	1	1

Input A true (+) makes all
outputs (B, C, D, E) true (+)

Figure 3–13. ENCODE gate symbol (Hewlett-Packard).

A (+) or (−) sign can be placed in the symbol to indicate the true state. In the circuit of Fig. 3–13, with A positive, all diodes conduct and all outputs are clamped positive. This would be true or 1 in a positive-true system. With A negative in the same system, all diodes would be open (nonconducting), producing a false or 0.

3–13. AMPLIFIERS, INVERTERS, AND PHASE SPLITTERS

When an amplifier is used in digital work (see Fig. 3–14), it is assumed that the output will be essentially the same as the input but in amplified form. That is, a true input will produce a true output, and vice versa. When inversion occurs, an inversion dot (or possibly an inverted pulse symbol) is placed at the output. Usually, the element is then called an *inverter* rather than an amplifier, even though amplification may occur. If a (+) or (−) sign is used in the symbol, this indicates the *input polarity* required to turn the amplifier on.

One amplifier or inverter symbol may represent any number of amplification stages, or optionally, separate symbols may be shown for each state.

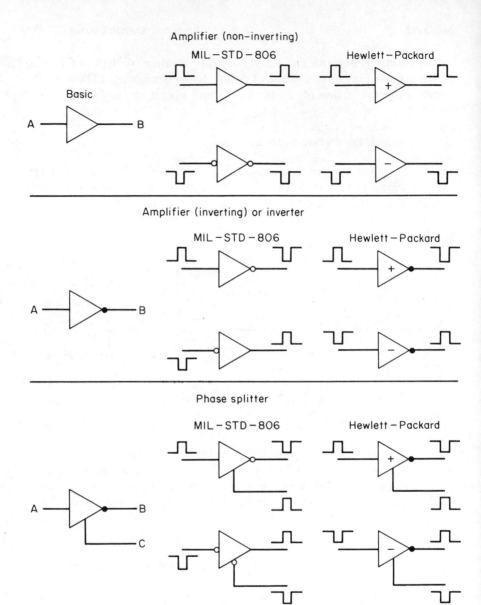

Figure 3–14. Amplifier, inverter, and phase-splitter symbols.

Individual logic symbols do not necessarily imply a specific number of components but rather relate to overall logic effect.

Sometimes an amplifier will be used as a *phase splitter* (one input and two outputs). One of the outputs is in phase with the input, while the other output is out of phase with the input. A similar case exists with *differential amplifiers,* which have dual inputs and dual outputs (although some differential amplifiers have dual inputs and a single output).

The rule for inversion dots on phase splitters and (particularly) differential amplifiers is that the dot indicates inversion of an output *with respect to the* corresponding input (not with respect to the opposite side of the amplifier). If an inversion dot is placed on the C output of the differential amplifier shown in Fig. 3–14, the C output is inverted with respect to the A input.

3–14. FLIP-FLOPS

A flip-flop (FF) is a bistable switching element; it takes one external signal to set the element and another signal to reset it. The FF will remain in a given state until switched to the opposite state by the appropriate external signal.

There are many types of FFs. Those used most frequently in digital equipment circuits are the reset-set (R-S), reset with clock, J-K, toggle, latching, and delay FFs.

3–14.1. Solid-State FF

A solid-state FF is shown in Fig. 3–15. Although this circuit uses discrete components, FFs are often found in IC form, with many FFs in one package. An example of this are *decade* FFs used in counting circuits, described in Sec. 3–20.

An FF has two stable states, with Q_1 conductive and Q_2 nonconductive, and vice versa. The first input pulse flips the circuit from one state to the other. The second input pulse flops the circuit back to its original state, hence the name flip-flop. Each time the circuit if flipped from one state to the other and back again (requiring two input pulses), a single (complete) output pulse is produced.

There are several methods for distinguishing between the two stable states of the FF. Generally, the start is to be considered with Q_1 cut off and Q_2 conducting. This is called the *0* state (or *false* state). The second state, with Q_1 conducting and Q_2 cut off, is called the *1* state (or *true* state).

Assume that in the circuit of Fig. 3–15, transistor Q_2 is conducting. The voltage drop across R_4 reduces the negative voltage at the collector of Q_2. This drop in negative voltage, coupled with the base of Q_1 through R_2 and

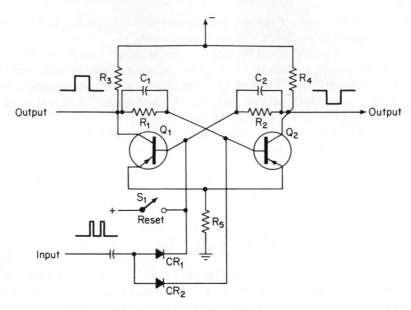

Figure 3–15. Solid-state FF circuit

C_2, reduces the forward base-emitter bias and drives Q_1 toward cutoff. (Both Q_1 and Q_2 are PNP. To be conductive, the base-emitter junction must be forward-biased. That is, the base must be negative relative to the emitter.) The negative voltage at the collector of Q_1 rises. This voltage rise, coupled with the base of Q_2 through R_1 and C_1, increases the forward base-emitter bias and drives Q_2 toward saturation. The process is cumulative and quickly results with Q_1 cut off and Q_2 conducting at saturation, the *0* state of the FF.

The FF may be flipped to its *1* state by driving Q_2 to cutoff. This may be done by applying a positive pulse of sufficient amplitude to its base. This results in Q_2 being cut off and Q_1 in saturation. The circuit may be flopped back to its *0* state by means of a second positive pulse applied to the base of Q_1.

The reset switch S_1 is a momentary switch which, when closed, places a positive bias upon the base of Q_1. This bias is sufficient to cut off Q_1 if Q_1 is in the conducting state. If Q_1 is already cut off, the bias has no effect. Thus, closing S_1 momentarily ensures that the cycle will start with the FF in its 0 state.

In a practical digital circuit, the reset switch S_1 is replaced by a repetitive pulse from a clock or time base. However, some electronic counters use a manual reset switch. The reset function is used primarily when the FF is used in a count/readout circuit rather than as a divider.

Output from the FF may be taken from the collector of Q_2. At the *0* state when Q_2 is at saturation, the collector voltage is some low negative value. At the *1* state, Q_2 is cut off and its collector voltage becomes more negative. Thus, the output for the entire cycle is a negative-going square pulse. If the output is taken from the collector of Q_1, the output is a positive-going square pulse.

3–14.2. FF Applications

The basic symbol and rules for application of FFs are shown in Fig. 3–16. The following notes provide an explanation of these rules.

The letters FF should appear in either the upper or the lower portion of the symbol thus identifying the element as an FF (rather than a one-shot multivibrator, Schmitt trigger, etc.).

An FF is assumed to be the simple R-S type if no other identification is made. When a *clock* input is added, the identifying letter C is placed inside the symbol. A clock input is usually a repetitive pulse (say, from a time base or programmer) and is *parallel connected to both* the set and reset side. Clock pulses are transient operated. That is, they are effective on leading or trailing edges of pulses, somewhat like an a-c coupled input.

If the clock input has no inversion dot, the input is effective on the true-going edge of the clock pulse. If an inversion dot is shown at the clock input, the clock input is effective on the false-going edge.

Multiple inputs on the *same side* of the FF symbol require the logical AND function (both inputs must be true to set or reset). Multiple inputs diagonally on the corner of the symbol require the logical OR function (either input true will set or reset). In some cases, a gate symbol (AND or OR) will be shown at the set or reset inputs. This is particularly true where multiple inputs are required.

3–14.3. R-S FF

The R-S FF has a minimum of two inputs, set and reset (A and B), and usually two outputs, set output and reset output (D and $\overline{D}$), as shown in Fig. 3–17.

The letter D indicates that the reset output, whether a 1 or a 0, is always the complement of the set output. That is, when D is true and $\overline{D}$ is false, the FF is defined as being in the set state. With D false and $\overline{D}$ true, the FF is in the reset state.

The FF is set by a true input to A (assuming no inversion dot on the symbol) and is reset by a true input to B. False inputs have no effect on a basic R-S.

Simultaneous true inputs to A and B are forbidden, since some intermediate output state would result. In practice, the FF would try to set and reset simultaneously, probably resulting in an undesired set or reset.

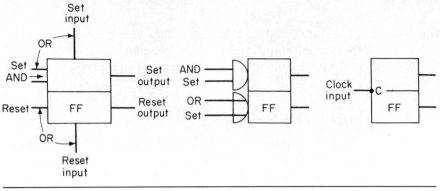

MIL – STD – 806

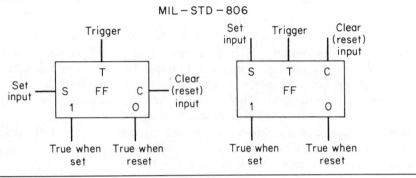

Hewlett – Packard

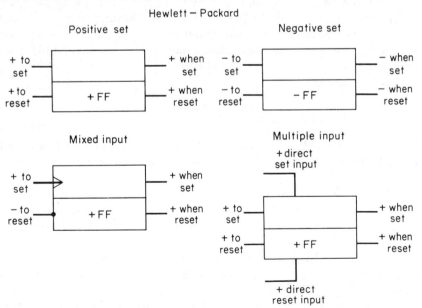

Figure 3–16. Basic FF symbols.

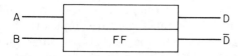

	Truth table		
A	B	D	$\overline{D}$
1	O	1	O
O	1	O	1
O	O	No change	

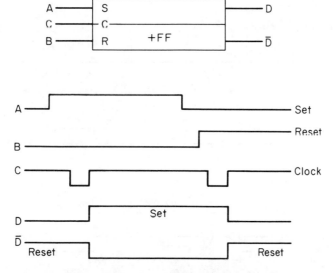

A ———

B ——— FF ——— D
              ——— $\overline{D}$

Figure 3–17. R-S FF symbol.

The truth table of Fig. 13–17 shows the three allowable input combinations for a basic R-S FF. If a-c coupled inputs are used, the FF will be set or reset by true-going transitions at A and B, respectively. If, in addition, input inversion dots are also used, false-going transitions at A or B will set or reset the FF.

3–14.4. R-S FF with Clock Input

An R-S FF with a clock (see Fig. 3–18) is similar to the basic R-S, except for the addition of the clock input (parallel connected to both set

Figure 3–18. R-S FF (with clock input) symbol.

and reset inputs.) A true input is required to both A and C to set the FF, and a true input to B and C is required for reset.

Since the clock input operates on a *pulse edge,* the setting or resetting must be present at A or B *before the clock pulse* transition occurs. This time relationship is shown (in positive-true form) in Fig. 13–18.

3–14.5. J-K FF

A J-K FF is used (instead of an R-S FF) where there is a possibility of two simultaneous true inputs which can result in an unpredictable output from an R-S FF. With a J-K FF, simultaneous true inputs for both set and reset will *reverse the existing state* of the FF. This requires some method of *storing* two information conditions (the existing output state and the new input state) until the clock pulse time.

Storage can be accomplished by (1) a-c coupling or (2) *dual-rank* FF, as shown in Fig. 3–19.

Truth table					
A B	Initial state		Resulting state	
	D	$\overline{D}$	D	$\overline{D}$
1 O			1	O
O 1			O	1
1 1	O	1	1	O
1 1	1	O	O	1
O O			No change	

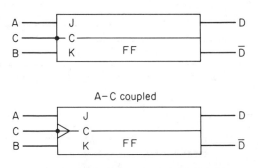

Figure 3–19. J-K FF symbol.

The a-c coupling method uses the RC time constant of the capacitive input for short-term storage of the input information. The dual-rank method combines two FFs (input storage and output storage) and several gates as a single logic element. For simplicity of representation, the internal dual-rank arrangement of the FF is not usually shown.

Overall operation of a J-K FF (either a-c coupled or dual-rank) is summarized as follows (as well as in the truth table of Fig. 3-19):

With true input at A only, the leading edge of the clock pulse acknowledges (stores) the information at A; the trailing edge of the clock pulse sets the FF.

With true input at B only, the leading edge of the clock pulse acknowledges (stores) the input information at B; the trailing edge of the clock pulse resets the FF.

With true inputs at both A and B, the leading edge of the clock pulse acknowledges the input information at A and B; the trailing edge of the clock pulse *switches the existing state* of the FF.

3-14.6. Toggle FF

The toggle FF has only one input, as shown in Fig. 3-20. Each time input A goes true, outputs D and $\overline{D}$ switch states. Since two input pulses or cycles are required to produce one complete cycle of the output, the toggle flip-flop acts as a *divide-by-two* element and is commonly used in a counting circuit (described in Sec. 3.20). The letter T inside the symbol identifies the toggle FF.

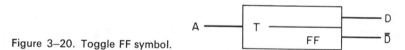

Figure 3-20. Toggle FF symbol.

3-14.7. Latching FF

The latching FF has a single signal input and a clock input. The symbol is identified by the letter L inside the symbol as shown in Fig. 3-21.

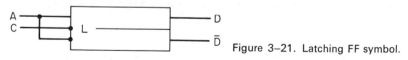

Figure 3-21. Latching FF symbol.

The set input responds to a true condition at A, while the reset input responds to a false condition at A. The FF will *latch* into the state existing at A (true-set or false-reset) when the clock pulse at C goes from true to false (usually at the trailing edge of the clock pulse). One unusual feature (not always desirable) is that during the clock pulse duration, the FF is *unlatched,* so that if A switches true and false several times during this period, outputs D and $\overline{D}$ will *free switch* accordingly.

3–14.8. Delay FF

The delay FF is similar to the latching FF, and its symbol is identified by the letter D as shown in Fig. 3–22. The delay FF will latch into the state existing at A when C goes from true to false. However, unlike the latching FF, the delay FF *delays* any switching of outputs until the trailing edge of the clock pulse occurs. The delay FF does not have an unlatched condition.

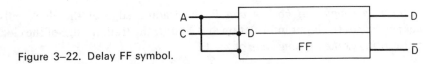

Figure 3–22. Delay FF symbol.

3–15. MULTIVIBRATORS

There are three types of multivibrators in general use with digital equipment: astable, one-shot (monostable), and Schmitt trigger (bistable).

3–15.1. One shot

The one-shot is a *monostable* switching element, using a multi-vibrator-type circuit. The one-shot element (sometimes known as a *mono*) is commonly used as an active delay device and is triggered into its unstable state by an external signal (see Fig. 3–23). After an interval determined by circuit constants, the one-shot returns automatically to the stable state. Thus, a known, fixed delay time is provided.

One-shot inputs are frequently a-c coupled, and triggering is accomplished when input A goes through a false-to-true transition. The abbreviation OS within the symbol identifies a one-shot, and a (+) or (−) sign may be used to indicate the true state of the D output during the *on* time.

3–15.2. Schmitt Trigger

The Schmitt trigger is a two-state (bistable) element, using a multivibrator-type circuit. The Schmitt trigger is commonly used for level sensing and signal squaring or shaping. When the input voltage is below a reference level, the element is in one state (see Fig. 3–24). When the input level goes above the reference level, the element switches to the other state. Switching between states takes place rapidly, making the element useful for squaring

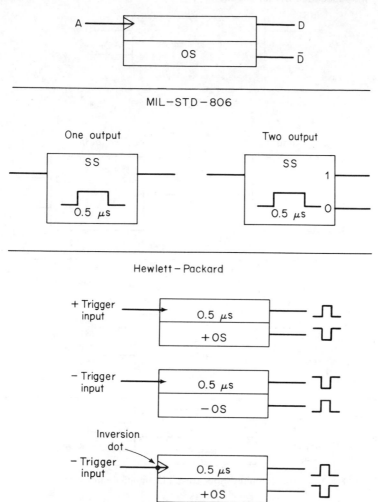

Figure 3–23. One-shot (monostable) multivibrator symbols.

signals with poor rise times (converting sinewaves into squarewaves) and for voltage-level restoration.

With input A below the reference level (or false), D is false and $\overline{D}$ is true. When the input is above the reference level, D switches to true and $\overline{D}$ switches to false. (The reference level is established by circuit constants.)

3–15.3. Multivibrator

Although all logic switching elements are forms of multivibrators, an *astable* type is assumed when the term multivibrator is used without a modifier (such as one-shot, Schmitt trigger, etc.). An astable multi-

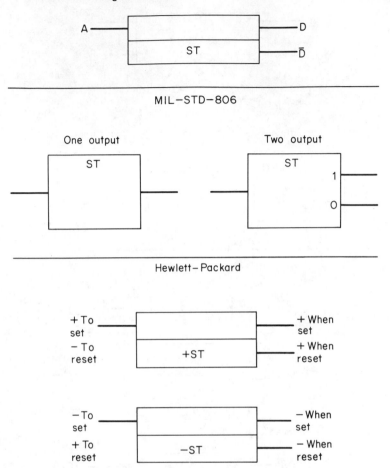

Figure 3–24. Schmitt trigger (bistable) multivibrator symbols.

vibrator will start a free-running operation when input A goes true and will continue to generate complementary pulse trains at outputs D and $\overline{D}$ until A goes false, as shown in Fig. 3–25. Typical negative-true timing waveforms are shown with the symbol. Note that the (−) sign preceding the letters MV indicates both the relative level required to start operation and the direction of the first output pulse at D. (The waveforms do not necessarily have to be symmetrical as shown.)

3–16. DELAY ELEMENTS

A delay element provides a finite time delay between the input and output signals. The delay symbols with examples of actual delay time are

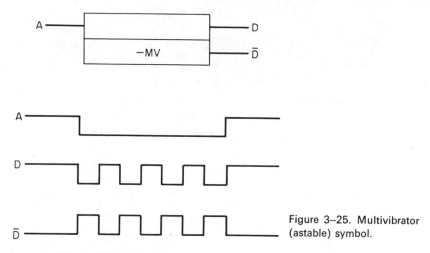

Figure 3–25. Multivibrator (astable) symbol.

shown in Fig. 3–26. Typical theoretical waveforms for the elements are shown adjacent to the symbols.

Many types of delay elements are used in digital work. Two frequently used delays are tapped delays and delays effective only on the leading or trailing edges of pulses. Such delay elements together with the theoretical waveforms are shown in Fig. 3–26.

3–17. LOGIC-SYMBOL IDENTIFICATION

Logic-symbol identification is one area where manufacturers take off in several directions at once. The following is a summary of the methods of logic-symbol identification in general use.

3–17.1. Reference Designations

Most logic diagrams show logic elements as a complete component rather than as the many components that make up the element. For example, an amplifier is shown as a triangle rather than as several dozen resistors, capacitors, transistors, and so on. Therefore, the amplifier has a reference designation of its own. On some logic diagrams, the logic-element symbols are mixed with symbols of individual resistors, capacitors, and so on. Either way, the logic-element symbol must be identified by a reference designation (to match descriptions in text or as a basis for parts listing).

The following logic-symbol reference designations are in general use:

Gates, G
Amplifiers and inverters, A

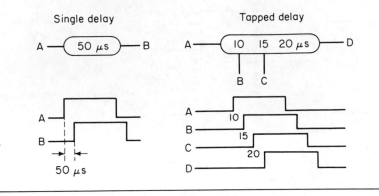

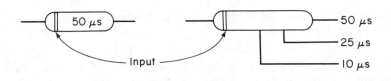

MIL – STD – 806

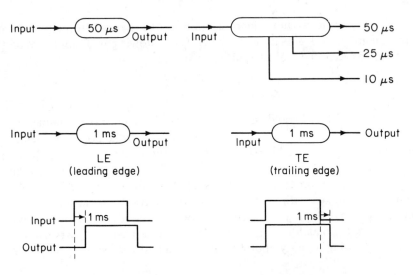

Hewlett – Packard

Figure 3–26. Delay-element symbols.

Flip-flops, FF
One-shot, OS (sometimes SS for single shot)
Multivibrator, MV
Schmitt trigger, ST
Delay, D.

Figure 3–27 shows some examples of how the reference designations are used. Note that the reference designations for switching elements are placed within the symbol. All other designations are (usually) placed beside the symbol. In the case of switching elements, the true-state sign can be used as a prefix to the designation.

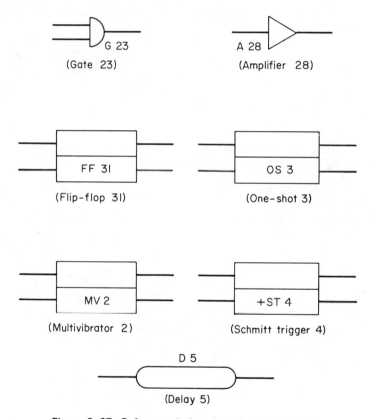

Figure 3–27. Reference designations for logic symbols.

3–17.2. IC Reference Designations

When logic elements appear in IC (or *microcircuit*) form, the system of reference designations is usually changed. Typical examples of microcircuit logic reference designations are shown in Fig. 3–28.

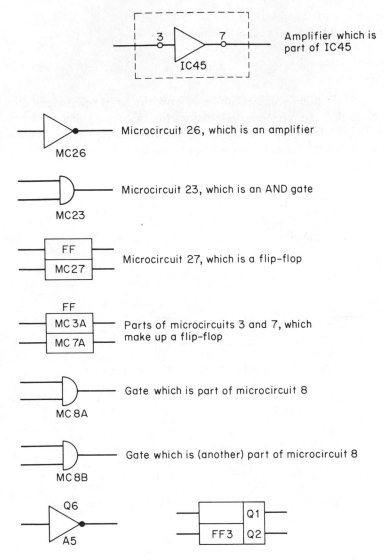

Figure 3–28. Reference designations for microcircuit (IC) logic symbols.

When a microcircuit is used to form one complete logic element, such as an amplifier or gate, the element can assume the microcircuit reference designator of MC rather than G or A.

When a microcircuit is used to form a complete switching circuit, the element can assume the microcircuit designator of MC, but it should also include the appropriate abbreviation (such as FF, OS, etc.) to identify its function.

When a switching element is composed of portions of different microcircuit packages, the designator MC of both packages should be included inside the symbol, and the appropriate identifying abbreviation should be located outside the symbol. When more than one logic element is included in the microcircuit package, each logic element is identified with a suffix: A, B, C, and so on.

On some diagrams, the logic element is shown enclosed by dotted lines with the terminals identified. An example of this is shown in Fig. 3–28 where an amplifier symbol is enclosed by dotted lines and identified as IC45. This indicates that the amplifier is part of IC45 (integrated circuit 45), that the input is available at terminal 3, and that the output is available at terminal 7.

On some logic diagrams which correlate to a schematic diagram of the same circuit, the *active components* of the element may also be designated. For example, in Fig. 3–28, Q_6 is the active transistor of amplifier A_5, and Q_1 and Q_2 are the active transistors of FF_3. This is primarily an aid in troubleshooting.

3–17.3. Reference Names

Identification of logic elements by functional name in a diagram is normally done only with switching elements, as shown in Fig. 3–29. The upper portion of the symbol can be reserved for this purpose. If the type of switching element (FF, OS, ect.) is not part of the designation (in the lower portion of the symbol) then the reference name should include the appropriate abbreviation. If designations appear in both the upper and lower portions of the symbol, the name can be placed outside the symbol. For one-shots, it is often convenient to give the *time duration* as the reference name.

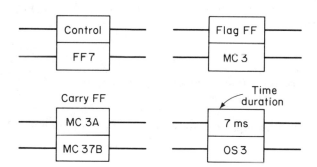

Figure 3–29. Reference names for logic symbols.

3–17.4. Location Information

Logic diagrams are intended to show the combination of logic elements that, taken together, form an instrument, part of an instrument, or a system of instruments. To aid in correlating the diagram with physical

locations in instruments, additional information is given (by some manufacturers) with logic symbols as shown in Fig. 3–30. The number in the small
triangle is sometimes used to indicate the circuit board on which the element
appears. In the example of Fig. 3–30, the elements appear on circuit board 2.

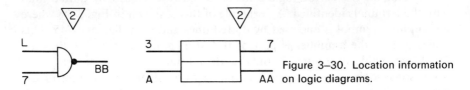

Figure 3–30. Location information on logic diagrams.

If all elements are on the same board, individual identification in this way
is unnecessary. Letters and numbers adjacent to inputs and outputs indicate
pin numbers of the board (or the microcircuit package) where inputs and/or
outputs appear, as shown in Fig. 3–28.

3–18. LOGIC EQUATIONS

Logic equations are sometimes used to aid in the explanation
of logic circuits. At one time, some manufacturers presented all logic information (in their instruction manuals) in equation form. Generally, this
practice has been discontinued. Logic diagrams are used in instruction
manuals. These diagrams may or may not be supplemented by the corresponding equations.

When logic equations are used, the familiar algebraic symbols have the
following meaning:

$+$ means *or*
$\cdot$ means *and*
$=$ means *equals* (or *is the result of*)
$-$ means *not* (or *complement of*)

For example, the logic equation for a simple two-input AND gate is

$$C = B \cdot A$$

The logic equation for the combination of elements shown in Fig. 3–31 is

$$E = (\overline{A} + B) \cdot (C + D)$$

Sometimes the same equation will be written as

$$(\overline{A} + B) \cdot (C + D) = E$$

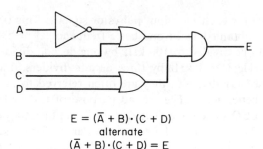

$$E = (\overline{A} + B) \cdot (C + D)$$
alternate
$$(\overline{A} + B) \cdot (C + D) = E$$

Figure 3–31. Logic diagram and
corresponding logic equations.

Generally, logic equations are better suited to *design* of digital equipment rather than as a tool for troubleshooting.

3–19. DIGITAL COMPONENT APPLICATIONS

The individual components described thus far can be arranged in various combinations to perform specific functions. For example, *decade counters* using FFs can convert a series of pulses into BCD form, such as conversion of a pulse count into an 8421 code. Also, a *decoder* can be used to convert the BCD data into a 10-line or decade readout.

Some other examples using combinations of basic digital components to perform specific functions include analog-to-digital conversion, digital-to-analog conversion, storage registers, shift registers, level comparators, sign or polarity comparators, adders, encoders, and a decoder. Operation of circuits that appear frequently in digital equipment are discussed in the following section.

3–20. COUNTER/READOUT AND DIVIDER CIRCUIT OPERATION

Most counters such as those used in digital readout equipment use decade counters that convert the count (series of pulses or events) into a BCD code and decoders for conversion of the code into decade form. In applications where the pulses or events must be read out, readout tubes are used to display the decade information.

The same circuits used for decade counters (four binary counters) are also used as dividers. Therefore, it is necessary to understand the operation of the basic decade before discussing how the decade is used in logic circuits.

3–20.1. Decade Circuits

Decade circuits serve two purposes in counters. First, the decade will divide frequencies by 10. That is, the decade will produce one output for each ten input pulses or signals. This permits several frequencies to be obtained from one basic frequency. For example, a 1-MHz time base can be divided to 100 kHz by one decade divider, to 10 kHz by two decade dividers, to 1 kHz by three decade dividers, and so on. When decades are used for division, they are often referred to as *scalers,* although *dividers* is a better term. The second purpose of a decade is to convert a count into a BCD logic code. The division function of a decade will be discussed first.

The basic unit of a decade divider is a 2-to-1 scaler, called a *binary counter.* This unit uses a *bistable multivibrator* or *an FF* circuit. The first input pulse flips the circuit from one state to the other. The second input pulse flops the circuit back to its original state. Each time the circuit is flipped from one state to the other and back again (requiring two input pulses), a single (complete) output pulse is produced.

The output pulses of one FF may be applied to the input of another similar FF for further frequency division. This is called *cascading.* A basic binary counter uses a cascaded chain of four FFs, as shown in Fig. 3–32. The count of this chain would be 16 (2^4).

Assume that all four FFs are in their 0 state, that a positive-going input is

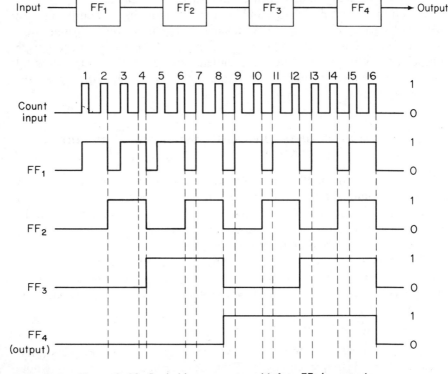

Figure 3–32. Basic binary counter with four FFs in cascade.

required to change FF states, and that the FF outputs are positive-going only when they shift from a 1 state to a 0 state.

Pulse 1 causes FF_1 to move from the 0 to the 1 state. The output of FF_1 is then negative-going and has no effect on FF_2. Pulse 2 causes FF_1 to move from the 1 to the 0 state. The output of FF_1 is then positive-going and moves FF_2 from the 0 to the 1 state. Pulse 3 causes FF_1 to move from the 0 to the 1 state (negative output with no effect on FF_2). Pulse 4 causes FF_1 to go from 1 to 0. This moves FF_2 from 1 back to 0. Thus, *four counts* are required for FF_2 to go through a *complete* cycle. This process is repeated, requiring 16 pulses before FF_4 will go through a complete cycle.

A *binary decade* counter uses feedback of the pulses to produce an output for ten input pulses, as shown in Fig 3–33.

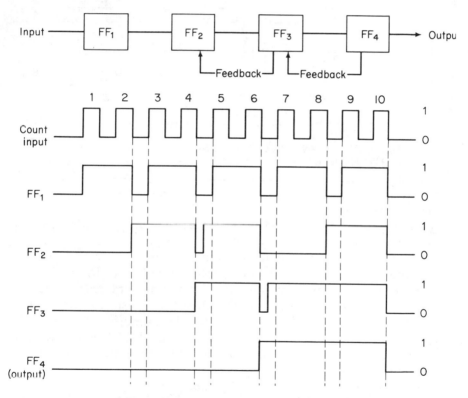

Figure 3–33. Binary decade counter with four FFs in cascade and feedback from FF_3 to FF_2 and FF_4 to FF_3.

3–20.2. Decade to BCD Conversion

Figure 3–34 shows a decade circuit capable of converting a series of pulses into an 8421 binary code. One FF is used for each of the

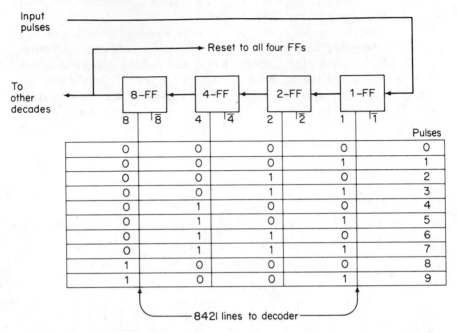

Figure 3–34. Binary decade circuit for converting a series of pulses into an 8421 binary code.

digits. Input pulses are fed into the 1-FF. The 2–, 4– and 8–FFs are cascaded, and receive pulses after the 1–FF.

For each state of an FF, one of the collectors is more positive than the opposite collector. When the states change, the polarities reverse. Some logic systems use *positive true,* while others use *negative true.* In the following discussions it is assumed that positive true is used for all circuits. That is, when an input or output is positive, that input or output is true. In the case of an FF, when one output or collector is true, the opposite collector is false.

For the FFs shown in Fig. 3–34 when the 8421 lines are positive, the 8421 lines are true. With the FFs in the *same state,* the $\overline{8421}$ lines are negative (false). When the FFs change states, the 8421 lines are negative (false) and the $\overline{8421}$ lines are true.

At the beginning of a count, the 8421 lines are at negative (false). This is represented by 0. The $\overline{8421}$ lines are positive (true). This is represented by 1. In a practical decade circuit the decades are set (or reset) to this condition by the application of a voltage or pulse. Typically, the pulses are supplied by a time base or programmer at the end of a 10 count.

When the first pulse in the count is applied, the 1–FF changes states. The 1 line becomes positive, represented by a 1 and the $\overline{1}$ line goes negative. (Note that the $\overline{8421}$ lines will always be at the *opposite* of the 8421 lines. The $\overline{8421}$ 0 and 1 states are omitted from Fig. 3–34 for clarity.)

When the second pulse is applied, the 1-FF changes states. The 1 line goes false (0) and the 2 lines goes true (1). With the third pulse applied, the 1–FF goes true (1) but the 2–FF remains true. Remember that the 2–FF will change states for each *complete cycle* of the 1-FF. When the fourth pulse is applied, the 1–FF goes to 0, as does the 2–FF. This causes the 4–FF to change states (the 4 line goes to positive or 1).

This process is repeated until a 9 count is reached. At that point, the 8–FF moves from 1 to 0 and produces an output. This output is returned to the reset line and serves to reset all of the FFs to the false state (8421 lines at 0). The output from the 8–FF can also be applied to the 1–FF of another decade. Any number of decades can be so connected. In a readout device, one decade is required for each readout tube.

3–20.3. BCD to Readout Conversion

Figure 3–35 shows the output of a decade connected to *decoder* which converts the BCD code into 10-line form. Each of the ten outputs (0 through 9) is connected to the corresponding cathodes of a readout tube. This is the basic circuit arrangement for a typical electronic counter. One readout tube (together with one decoder and one decade) is required for each digit of the counter. In a practical circuit, a *register* and/or *storage circuit* is placed between the decades and their corresponding decoders. These registers will hold the BCD count data until it is cleared by the operator or will sample the count at some given rate (usually two or three times per second). Operation of registers is discussed in Sects. 3–22 and 3–23.

Note that the decoder is essentially a group of AND gates connected to provide nine inputs (8421, $\overline{8421}$, and an *enable* input) and ten outputs (0 through 9). As discussed previously, an AND gate requires that all inputs be true (positive in this case) to produce an output. All of the AND gates in Fig. 3–35 have three inputs which must be true before they will produce an output.

The output of gates 0 through 9 are connected to the cathodes of a readout tube. Each readout tube consists of ten cold cathodes and a common anode, all enclosed in a gas-filled envelope. These cathodes are shaped as numbers 0 to 9 and are stacked one above the other. When one of the decoder gates (0 through 9) produces an output, the corresponding cathode circuit is completed. The gas between that particular number-shaped cathode and the common anode is ionized, causing the cathode to glow, thereby revealing its number at the front panel. Only one decoder gate produces an output at any given time. Thus, the remaining cathodes are non-ionized and remain unlit (not visible from the front panel).

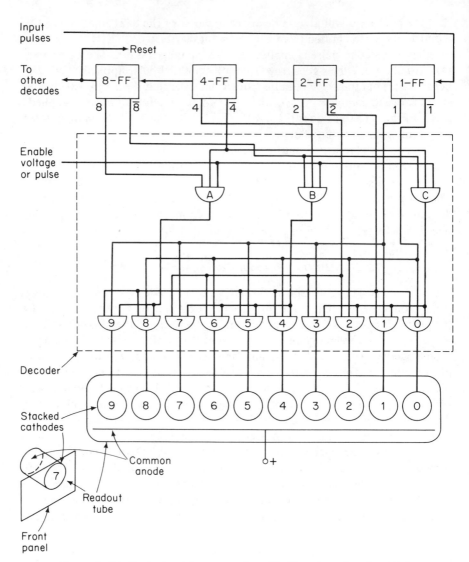

Figure 3–35. Basic circuit for conversion of BCD code to decade readout.

Assume that the circuit of Fig. 3–35 is part of an electronic counter, and the the counter's gate is held open for a count of seven input pulses. Under these conditions, the $\overline{8}$, 4, 2, and 1 lines would be true (1). Likewise, the 8, $\overline{4}$, $\overline{2}$, and $\overline{1}$ lines would be false (0). The B AND gate of the decoder would have three true inputs ($\overline{8}$, 4, and enable). The enable pulse or voltage is always true and is applied (manually or by the sample rate generator) whenever a readout is desired.

Both the A and C gates will have at least one false input. Therefore, only the 4, 5, 6, or 7 gates could produce an output. The 7 gate would have three true inputs (B gate, 1 line, and 2 line). Gates 4, 5, and 6 would have at least one false input. Therefore, only the 7 gate will produce an output, and the number 7 cathode of the readout tube will glow.

3–21. CONVERSION BETWEEN ANALOG AND DIGITAL INFORMATION

There are several methods for converting voltage (or current) into digital form. One method is to convert the voltage into a frequency or series of pulses. Then the pulses are converted into BCD form, and the BCD data is converted into a decade readout.

It is also possible to convert voltage directly into BCD form by means of an analog-to-digital (a/d) converter and to convert BCD data back into voltage by means of a digital-to-analog (d/a) converter.

Before going into operation of these conversion circuits, let us discuss the *signal formats* for BCD data as well as the four-bit system.

3–21.1. Typical BCD Signal Formats

Although there are many ways in which pulsed waveforms can be used to represent the 1 and 0 digits of a BCD code, there are only three ways in common use. These are the NRZL (nonreturn-to-zero-level), the

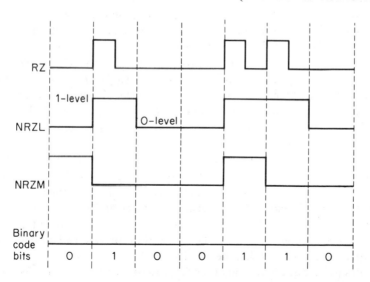

Figure 3–36. Typical BCD signal formats.

NRZM (nonreturn-to-zero = mark), and the RZ (return-to-zero) formats. Fig. 3–36 shows the relation of the three formats.

In NRZL, a 1-bit is one signal level, while a 0-bit is another signal level. These levels could be 5 V, 10 V, 0 V, -5 V, or any other selected values, provided the 1 and 0 levels are *entirely different* and predetermined.

In NRZM, the level of the waveform has no meaning. A 1–bit is represented by *change in level* (either or lower), while a 0-bit is represented by *no change in level*.

In RZ, a 1-bit is represented by a pulse of some definite width (usually a $\frac{1}{2}$-bit width) that returns to zero signal level, while a 0-bit is represented by a zero-level signal).

3–21.2. The Four Bit System

As the name implies, a four-bit system is one that is capable of handling four information bits. Any fractional part *in sixteenths* can be stated by using only four binary digits. As discussed, the number 15 is represented by 1111 in the binary system. (Zero is then represented by 0000.) Therefore, any number between 0 and 15 requires only four binary digits. For example, the number 1101 is 13, the number 0011 is 3, and so on. Although not all BCD codes used the four-bit system, it is common and does provide a high degree of accuracy for conversion between analog and digital data.

In practice, a four-bit a/d converter (sometimes known as a binary encoder) samples the voltage level to be converted and compares the voltage to $\frac{1}{2}$-scale, $\frac{1}{4}$-scale, $\frac{1}{8}$-scale, and $\frac{1}{16}$-scale (in that order) of some given full-scale voltage. The encoder then produces bits in sequence, with the decision made on the most significant (or $\frac{1}{2}$-scale) first.

Figure 3–37 shows the relation between three voltage levels to be encoded (or converted) and the corresponding binary code (in NRZL form). As shown, each of three voltage levels is divided into four, equal, time increments. The first time increment is used to represent the $\frac{1}{2}$-scale bit, the second increment is used for the $\frac{1}{4}$-scale bit, the third increment is used for the $\frac{1}{8}$-scale bit, and the fourth increment is for the $\frac{1}{16}$-scale bit.

In level 1, the first two time increments are at a binary 1, while the second two increments are at a binary 0. This would be represented as 1100, or 12. Twelve is three-fourths of 16. Therefore, level 1 is 75 per cent of full scale. For example, if full scale is 100 V, level 1 would be at 75 V.

In level 2, the first two time increments are at a binary 0, while the second two increments are at a binary 1. This would be represented as 0011, or 3. Therefore, level 3 is $\frac{3}{16}$ of full scale (or. 18.75 V).

This could be expressed in another way. In the first or $\frac{1}{2}$-scale increment the encoder produces a binary 0. This is because the voltage (18.75 V) is less

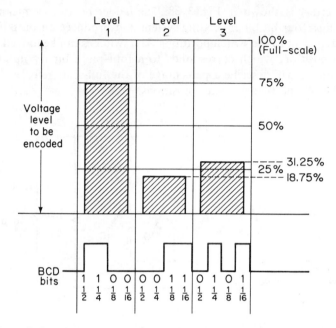

Figure 3–37. Relationship between three voltage levels to be encoded and corresponding BCD code (using four-bit system).

than $\frac{1}{2}$-scale (50 V). The same is true of the second or $\frac{1}{4}$-scale increment (18.75 V is less than 25 V). In the third or $\frac{1}{8}$-scale increment, the encoder produces a binary 1, as it does in the fourth or $\frac{1}{16}$-scale increment. This is because the voltage being compared is greater than $\frac{1}{8}$ of full scale (18.75 is greater than 12.5) and greater than $\frac{1}{16}$ of full scale (18.75 is greater than 6.25).

Therefore, the one-half and one-quarter increments are at zero or *off*, while the $\frac{1}{8}$- and $\frac{1}{16}$-scale increments are *on*. Also, $\frac{1}{8} + \frac{1}{16} = \frac{3}{16}$, or 18.75 per cent.

In level 3, the first time increment is 0, the second time increment is 1, the third is 0, while the fourth is 1. This is a binary 0101, or 5, and represents $\frac{5}{16}$ of full scale (31.25 V).

3–21.3. A/D Conversion

One of the most common methods of *direct* a/d conversion involves the use of an encoder which uses a sequence of half-split, trial-and-error steps. This produces code bits in *serial* (all on one line, one after another) form.

The heart of an encoder is a *conversion ladder*. Such a ladder (a form of a/d converter) is shown in Fig. 3–38. The ladder provides a means of implementing a four-bit binary coding system and produces an output voltage that is equivalent to the switch positions. The switches can be moved to either a 1 or 0 position, which corresponds to a four-place binary number. The output voltage will describe a percentage of the full-scale reference voltage, depending on the binary switch positions.

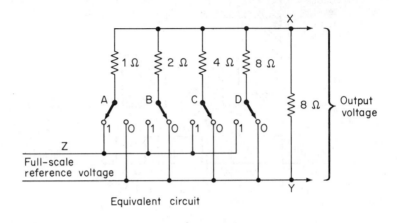

Equivalent circuit

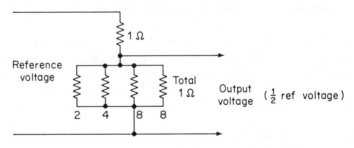

Figure 3–38. Binary conversion ladder used in four-bit system.

For example, if all of the switches are in the 0 position, there is no output voltage. This produces a binary 0000 (conventional zero) represented by zero voltage.

If switch A is in the 1 position and the remaining switches are in the 0 position, this produces a binary 1000 (conventional 8). Since the total in a four-bit system is 16 (as described previously), this should represent one-half (8 = one-half of 16). Therefore, the output voltage should be one-half of the full-scale reference voltage. This is accomplished as follows.

The 2-, 4- and 8-Ω switch resistors and the 8-Ω output resistor are connected in parallel. This produces a value of 1 Ω across points X and Y. The reference voltage is applied across the 1-Ω switch resistor (across points Z and X) and the 1-Ω combination of resistors (across points X and Y). In effect, this is the same as two 1-Ω resistors in series. Since the full-scale reference voltage is applied across both resistors in series and the output is measured across only one of the resistors, the output voltage will be one-half of the reference voltage.

3–21.3.1. Practical Encoder

In a practical encoder circuit, the same basic ladder is used to supply a comparison voltage to a *comparison circuit*. This circuit compares the voltage to be encoded against the binary-coded voltage from the ladder. The resultant output of the comparison circuit is a binary code representing the voltage to be encoded.

The mechanical switches shown in Fig. 3–38 are replaced by electronic switches, usually FFs. When the FF is in one state (the *on* state), the corresponding ladder resistor is connected to the reference voltage. In the *off* state, the resistor is disconnected from the reference voltage. The switches are triggered by four pulses (representing each of the four binary bits) from the time base (clock or programmer). A fifth pulse is used to turn the comparison circuit on and off, so that as each switch is operated, a comparison can be made for each of the four bits.

Figure 3–39 is a simplified block diagram of such an encoder. Here, the reference voltage (equivalent to a full-scale input voltage or voltage to be encoded) is applied to the ladder through the electronic switches. The ladder output (comparison voltage) is controlled by switch positions which in turn are controlled by pulses from the time base. The sequence of these pulses is also shown in Fig. 3–39.

The following paragraphs describe the sequence of events necessary to produce a series of four binary bits that describe the input voltage as a percentage of full scale (in one-sixteenth increments). Assume that the input voltage is three-fourths full scale.

When pulse 1 arrives, switch 1 is turned on and the remaining switches are turned off. The ladder output is a $\frac{1}{2}$-scale voltage that is applied to the differential amplifier. The balance of this amplifier is set so that its output will be sufficient to turn on one AND gate and turn off the other AND gate if the ladder voltage is greater than the input voltage. Likewise, the differential amplifier will reverse the AND gates if the ladder voltage is not greater than the input voltage. Both AND gates are placed in such a way as to turn on (the gates are *enabled*) by the fifth pulse from the time base.

In our sample (three-fourths full scale), the ladder output is less than the input voltage when pulse 1 is applied to the ladder. As a result, the *not greater* AND gate will turn on, and the output FF will be set to the 1 position. Therefore, for the first of the four bits, the FF output is a 1.

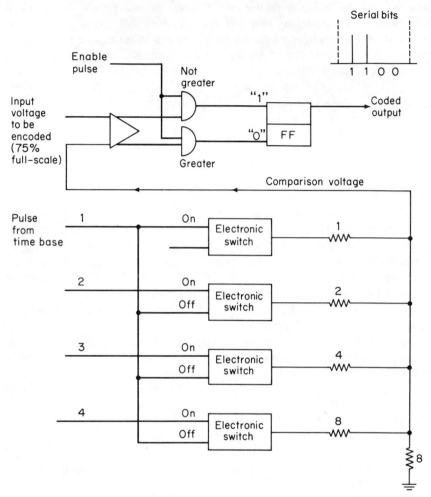

Figure 3–39. Analog-to-digital converter using four-bit system.

When pulse 2 arrives, switch 2 is turned on and switch 1 remains on. Both switches 3 and 4 remain off. The ladder output is now a $\frac{3}{4}$-scale voltage that is applied to the differential amplifier. Therefore, the ladder voltage equals the input voltage. However, the ladder output is still *not greater* than the input voltage. Consequently, when the AND gates are enabled by the fifth pulse, the AND gates remain in the same condition as does the output FF.

When pulse 3 arrives, switch 3 is turned on. Switches 1 and 2 remain *on,* while switch 4 is *off*. The ladder output is now a $\frac{7}{8}$-scale voltage that is applied to the differential amplifier. The ladder output is now greater than the input voltage. Consequently, when the AND gates are enabled, they reverse.

The not greater AND gate turns off and the *greater* AND gate turns on. The output FF then sets to the 0 position.

When pulse 4 arrives, switch 4 is turned on. All switches are now on. The ladder output is now maximum (full scale). This output is applied to the differential amplifier. The ladder voltage is still greater than the input voltage. Consequently, when the AND gates are enabled, they remain in the same condition as does the output FF.

Therefore, the four binary bits from the output are 1, 1, 0, and 0, or 1100. This is a binary 12, which is three-fourths of 16.

In a practical encoder, when the fourth pulse has passed, all of the switches are reset to the off position. This places them in a condition to spell out the next four-bit binary word.

3–21.4. D/A Conversion

A d/a converter performs the opposite function of the a/d converter just described. A d/a converter produces an output voltage (usually dc) that corresponds to the binary code. Information to be converted is usually applied to the d/a input in serial form. However, some d/a converters receive 4-line data.

As shown in Fig. 3–40, a conversion ladder is also used in the d/a converter. The output of the conversion ladder is a d-c voltage that represents a percentage of the full-scale reference voltage. The output percentage depends on the switch positions. In turn, the switches are set to *on* or *off* by corresponding binary pulses. If the information is applied to the switches in 4-line form, each line can be connected to the corresponding switch. If the data is applied

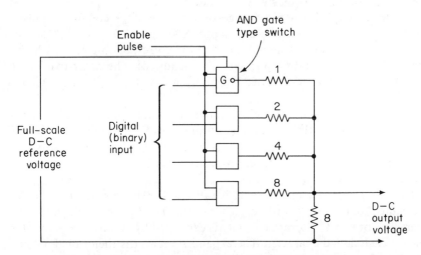

Figure 3–40. Digital-to-analog converter using four-bit system.

in serial form, the data must be converted to parallel (4-line) form by a shift register and/or storage register (described in Secs. 3–22 and 3–23).

The switches in a d/a converter are essentially a form of AND gate. Each gate completes the circuit from the reference voltage to the corresponding ladder resistor when *both* the enable pulse and the binary pulse coincide.

Assume that the digital number to be converted is 1000 (a binary 8). When the first pulse is applied to the ladder switches, switch A is enabled and the reference voltage is applied to the 1-Ω resistor. When switches B, C, and D receive their enable pulses, there will be no binary pulses (or they will be in their low-level 0 condition, depending on the binary pulse format used). Therefore, switches B, C, and D will not complete the circuits to the 2-, 4- and 8-Ω ladder resistors. These resistors combine with the 8-Ω output resistor to produce a 1-Ω resistance in series with the 1-Ω ladder resistance. This divides the reference voltage in half to produce a $\frac{1}{2}$-scale output.

3–22. STORAGE REGISTERS

The storage register is a combination of gates and FFs used to store binary information. A typical storage register is illustrated in Fig. 3–41. This register can store four bits of binary information applied to inputs A, B, C, and D. The inputs must be positive true.

Assume that A and C are at positive levels (true) and that B and C are negative (false).

Before data are stored, all four FFs are reset by the *leading edge* of the reset/store command pulse (from the time base, clock, or programmer). This clears the register of any previously stored data.

The trailing (positive-going) edge of the reset/store command pulse is coupled into the four positive-true input AND gates. Inputs that are true (A and C) will enable a true output from the corresponding gates, which in turn will set FFs A and C (positive input required). The other two FFs (B and D) remain in the reset state. The information at A, B, C, and D is now stored in the four FFs.

When the stored data is required, the output enable line is made negative (by a pulse from the time base), and output lines E and G from the set FFs are true (negative true, now). Outputs F and H will be false (positive). (If positive-true outputs are required from the FFs, then the reset output of the FFs can be used.)

The storage register of Fig. 3–41 is called a 4 × 1 *register*. The 4 denotes the number of bits per "word" or "character." The 1 indicates the number of characters or words. Storage capacity is expanded by adding 4 × 1 registers, one for each added character. The expanded register is then called a 4 × 2, 4 × 3, and so on register. If the characters require more than four bits, the

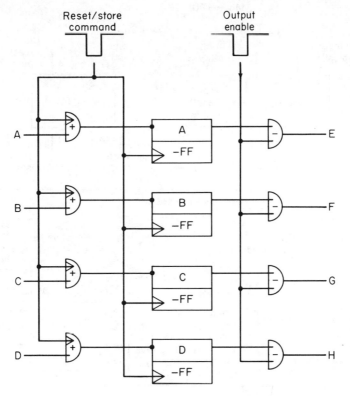

Figure 3–41. Typical storage register.

register is increased by 5 × 1, 5 × 2, and so on by adding storage FFs and associated gates (see Fig. 3–42).

By using one store pulse, parallel data (all four binary pulses occurring simultaneously) are stored. Then by successively enabling the register for each character, parallel-to-serial conversion is obtained (corresponding bits of each register are OR gated together). The process of successively enabling a character storage register is sometimes called *commutation*.

3–23. SHIFT REGISTERS

Shift registers have many functions. A shift register is sometimes used to perform arithmetic functions of multiplication and division. When a number is multiplied by 10, it is shifted to the left and a zero added. For example, the number 377 × 10 is 3770 (377 shifted to the left and 0 added). Division, the opposite of multiplication, requires a right shift. A shift register is also used for serial-to-parallel conversion of binary pulses.

MIL−STD−806

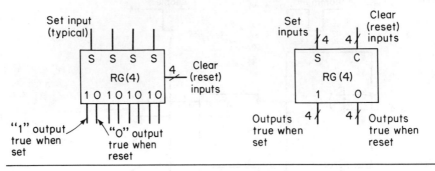

Hewlett−Packard

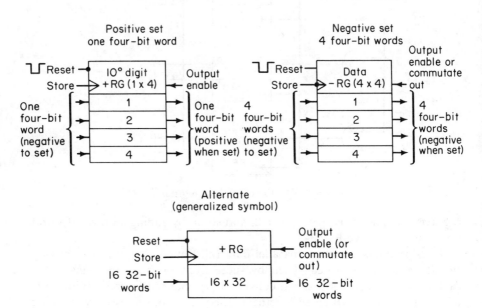

Figure 3–42. Typical storage-register symbols.

As shown in Fig. 3–43, a shift register is made up of FFs and delay elements. The FFs are connected together so that each one transfers its existing data to the next FF when the reset input is applied. (The reset input pulse applied simultaneously to all FFs is often known as the *advance* input when referring to a shift register.) One FF is required for each bit to be converted; thus four FFs are shown for four binary bits. This is similar but not identical to the decade discussed previously.

When the advance (reset) line is pulsed as it is between every binary pulse bit, the FFs in the set (1) state switch to 0, and those in the reset (0) state

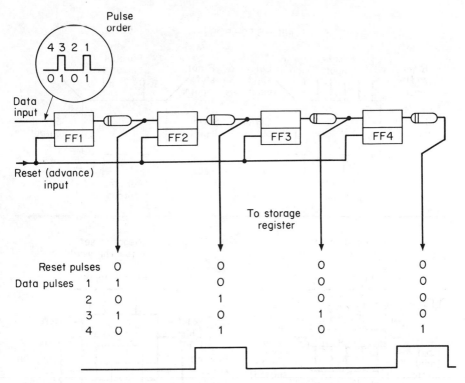

Figure 3–43. Typical shift register.

remain in the 0 state. The output signal from a switching FF is shaped and delayed in the delay circuits and then is directed to the 1 input of the next FF to complete the shift operation. Data can be inserted into the first FF between each shift.

Assume that four binary bits in serial form are applied to the input of the register in Fig. 3–43 and that the pulses appear as 0101 (a binary 5) as shown. All of the FFs are in the 0 state before each binary pulse. At pulse 1, FF_1 goes to the 1 state and the remaining FFs stay in the 0 state. Between pulses 1 and 2, all FFs go to 0 but the 1 condition of FF_1 is delayed and applied to FF_2.

At pulse 2, FF_1 stays in the 0 condition (since the second bit is 0) but FF_2 moves to a 1 state, since it received the delayed pulse from FF_1. At pulse 3, FF_1 goes to the 1 state, FF_2 goes to 0, and FF_3 now goes to 1, since it received the delayed pulse from FF_2. At pulse 4, FF_1 remains in the 0 state, FF_2 moves to the 1 state (pulse from FF_1), FF_3 goes to 0 (since it receives a 0 from FF_2), and FF_4 (thus far unaffected) moves to a 1 state (received from FF_3).

MIL – STD – 806

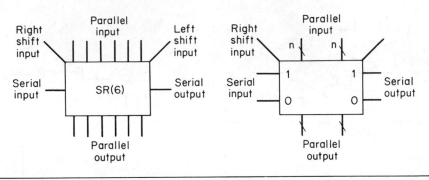

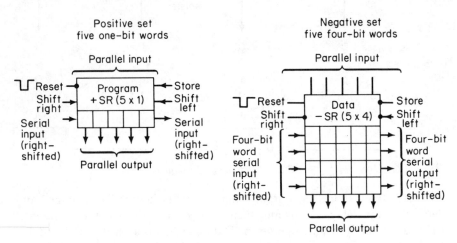

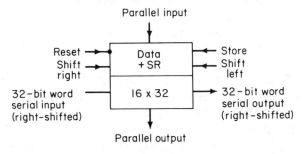

Figure 3–44. Typical shift-register symbols.

At the end of the four bits, before all FFs are reset to 0, the existing states (0101) are fed into a storage register. Then the output of the storage register can be used as needed, for example, to operate the electronic switches of a d/a converter.

As shown in Fig. 3–44, the shift register is usually presented in symbol form on logic diagrams.

3–24. SIGN COMPARATORS

Sign comparators have many uses in digital circuits. Typically, a sign comparator is used to determine the sign of numbers (plus or minus) or to compare the polarity of d-c voltages.

As shown in Fig. 3–45, a sign comparator consists of two FFs, four AND gates, and two OR gates. The two FFs, each indicating the polarity of an applied voltage, can produce four possible combinations, two for like polarities and two for unlike polarities.

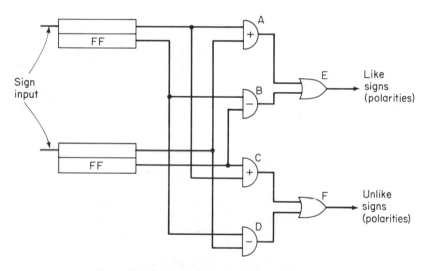

Figure 3–45. Typical sign (polarity) comparator.

If the FFs have the same output, one of the AND gates (A for positive or B for negative) will have an output, and OR gate E will provide an output signal. That is, there will be an output indicating *like* polarities, whether both FFs have a positive or negative output. One of the two AND gates will provide an output signal, and since an OR gate requires only one input to produce an output, OR gate E will provide a *like* polarity output.

If the two polarities are opposites (one positive and one negative) there are only two possible combinations. For either combination, one of two AND gates (C or D) will provide an output to OR gate F, indicating an *unlike* polarity output.

3–25. ADDERS

Adders (half-adders and full-adders) are used primarily in computer circuits to perform mathematical functions. However, adders can also be used as some form of comparison circuit.

3–25.1. Half-Adder

When only two binary digits are to be added, a half-adder can be used. The output will be a sum $(1+0=1, 0+1=1)$ or a carry $(1+1=10,$ which is a 0 sum and carry 1, or a binary 2). The two digits to be added have only four possibilities, as shown in Fig. 3–46. The digit be to added (A) is the *addend* and the other digit (B) is the *augend*.

Addend (digit A)	Augend (digit B)	Sum	Carry
0	0	0	0
0	1	1	0
1	0	1	0
1	1	0	1

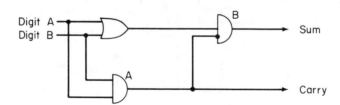

Figure 3–46. Typical half-adder.

If both A and B digits are 0 (or false) there will be no output from the OR gate. Likewise, there will be no output from AND gate A; hence, no carry. That is, the output from AND gate A will be 0 (false).

Note that this same carry output is also applied to AND gate B through an inversion. Therefore, if the carry output is 0, the inverted input to AND gate B will be 1. With a 0 from the OR gate and a 1 from the inverted input, there can be no output from AND gate B. Therefore, we have 0 sum and 0 carry.

If either A or B digits equal 1, there will be an output from the OR gate to AND gate B, but the AND gate A output will remain 0. This 0 output is inverted to a 1 and is applied to AND gate B. Now, a 1 from the OR gate and a 1 from the inversion result in a sum output of 1 and a 0-carry output.

If both A and B digits equal 1, they will pass through the OR gate to produce a 1 and also through the AND gate to A to produce a carry output of 1. This carry 1 is inverted to AND gate B, so a 1 and a 0 are applied to AND gate B, and there is no sum output. Therefore, the sum is 0, and the carry is 1.

3–25.2. Full Adder

A full-adder circuit is shown in Fig. 3–47. This circuit makes it possible to add two three-digit binary numbers. As shown, there are eight

	A	B	C		
	Addend	Augend	Carry (in)	Sum	Carry (out)
1	1	0	0	1	0
2	0	1	0	1	0
3	0	0	1	1	0
4	1	1	0	0	1
5	1	0	1	0	1
6	0	1	1	0	1
7	1	1	1	1	1
8	0	0	0	0	0

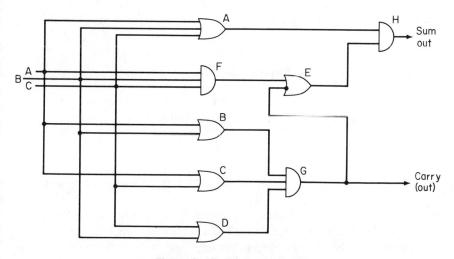

Figure 3–47. Typical full-adder.

possible combinations. However, aside from all zeros, there are only three different cases, depending on the number of 1s (trues—either one, two, or three 1s in any combination.

For a single 1, there are three possibilities: $(1+0+0)$, $(0+1+0)$, and $(0+0+1)$. Thus, there is always a sum but never a carry. For any two 1s $(1+1+0, 1+0+1, \text{ and } 0+1+1)$ there is a carry but never a sum. For three 1s $(1+1+1)$, there are both a carry and a sum. Where every digit is zero $(0+0+0)$ there is neither a sum nor a carry.

If all digits are 1s, the three OR gates B, C, and D each have two 1s. So there is a 1 from each, resulting in three 1s in AND gate G. This provides a carry, which is inverted and appears as a 0 at OR gate E. But the three 1s into AND gate F provide a 1 input to OR gate E; hence, OR gate E has an output. OR gate A has three 1s, so there are two 1s input at AND gate H. This produces both a sum and a carry output $(1+1+1=11, \text{ a binary } 3)$.

If two digits are 1s, the three OR gates B, C, and D will each have at least one 1. AND gate G will have three 1s, providing a carry output. AND gate F will have two 1s and no output (0 output). AND gate F will have an output (carry), but it will be inverted. Therefore, both inputs to OR gate E will be 0, as will its output to AND gate H. This produces a carry but no sum $(1+1+0=10, 1+0+1=10, 0+1+1=10)$.

If only 1 digit is a 1, there cannot be three inputs to AND gate G; hence, no carry. The inversion will change this carry 0 to a 1 which OR gate E will pass to AND gate H. Another input from OR gate A is applied to AND gate H, producing a sum output. This results in a sum but no carry $(1+0+0 =1, 0+1+0=1, 0+0+1=1)$.

If all digits are 0, there will be no input to AND gate G; hence, no carry. There will be no output for OR gate A; hence, there is no output from AND gate H and no sum $(0+0+0)$.

3–26. COMPARISON CIRCUIT

The half-adder can be used as a comparison circuit for detecting errors. If two unlike digits are fed into the half-adder, the output will be a sum. If the digits are alike (0, 0 or 1, 1) there will be no sum output (even though there may be a carry output). Therefore, the half-adder is a detector of equality or inequality.

Figure 3–48 shows how a half-adder is used to compare the contents of two registers. The binary count in register A is transferred to register B, and the half-adder must make sure that both registers contain the same count. The half-adder compares the contents of A and B registers, digit by digit. As long as the half-adder has no sum output, the digit in A is the same as its respective digit in B. If there is a sum output, the digit in A is unlike the digit in B.

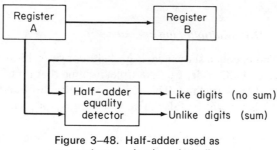

Figure 3–48. Half-adder used as comparison circuit (equality detector).

3–27. MULTIPLE-LINE ENCODERS AND DECODERS

In digital equipment, it is often necessary to convert BCD data into multiple-line form, and vice versa. When multiple-line data is converted to BCD, the function is known as encoding and requires the use of an *encoder* (sometimes known as a matrix). A decoder is the opposite of an encoder. That is, the decoder converts BCD data into multiple-line form. (An example of this is discussed in Sec. 3–20).

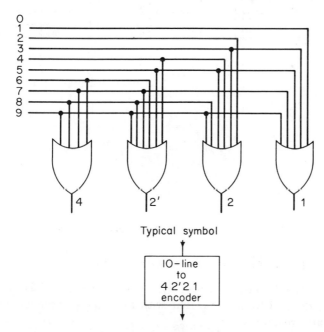

Figure 3–49. Typical multiple-line-to-BCD encoder.

3-27.1. Multiple-Line Encoder

The encoder shown in Fig. 3-49 converts 10-line decimal information to 42′21 BCD code. For example, assume that input line 3 is true,
but all other input lines are false. This will make OR gates 1 and 2 true, while
the 4 and 2′ OR gates will be false. The resultant output will be 0011, or a
binary 3.

An encoder may have any number of inputs (only one of which is true at
one time) and any number of binary bits on the output (various combinations
of which may be made true). Encoders are often shown simplified on logic
diagrams as a rectangular box with the type of matrix (10-line to 42′21,
etc.) indicated within the symbol.

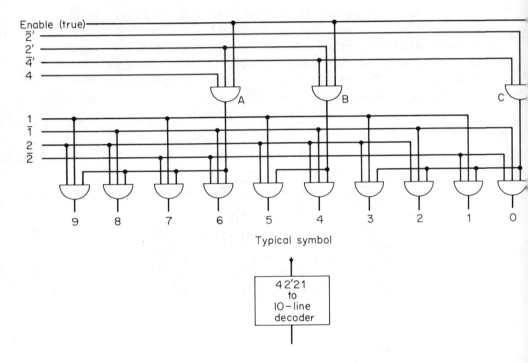

Figure 3-50. Typical BCD to multiple-line decoder.

3-27.2. Multiple-Line Decoder

The decoder shown in Fig. 3-50 converts a 42′21 BCD code
into 10-line decimal form. Note that with this type of decoder, the $\overline{42'22}$
inputs are also required.

Assume that the BCD input in 7 (1101 for a 42'21 code). Under these conditions, gate A will be true, but gates B and C will be false (the $\overline{4}$ input line will be false). Therefore, the 6, 7, 8, and 9 output gates could be true, but the 0, 1, 2, 3, 4 and 5 output gates must be false. The 7 output gate has all three inputs true (from gate A, the 1 line and the $\overline{2}$ line) and therefore produces a true output. Gates 6, 8, and 9 have at least one false input and thus produce false outputs.

Decoders are sometimes shown simplified on logic diagrams as a rectangular box with the logic arrangement (42'21 to 10–line, etc.) indicated within the symbol.

4. BASIC DIGITAL
TROUBLESHOOTING

Chapter 3 introduced digital equipment circuits. This chapter covers those basic troubleshooting techniques that apply to digital circuits. Although the theoretical and practical principles of troubleshooting and service, such as those described in Chapters 1 and 2, are used for digital equipment, there are many practical techniques that apply only to digital equipment.

As one example, the key technique described here is based on measurement of pulses at inputs and outputs of digital circuits. From a troubleshooting standpoint, a complete digital logic diagram can appear as a hopeless maze, particularly if the logic circuits are *interlaced* (that is, multiple functions for the same signal at different times, or one signal depending on many other signals or signal conditions). However, if pulses can be checked at the input and/or output of each circuit or group of circuits, the condition of that circuit can be checked quickly.

This troubleshooting technique requires the use of an oscilloscope and possibly a pulse generator. Such a combination of test equipment will quickly determine the presence or absence of pulses at circuit inputs and outputs, as well as the pulse duration, amplitude, and delay. For example, an AND gate may show an incorrect output even though there are two input pulses of correct amplitude and duration, if one pulse is delayed.

These measuring and monitoring techniques together with many other test and fault localization methods for digital circuits are discussed in the following sections. It is recommended that the reader make a thorough study of the specialized repair methods for ICs discussed in Chapter 2. Much of today's digital logic is found in IC form.

4–1. DIGITAL TEST EQUIPMENT

The test equipment and tools required for service of digital equipment are essentially the same as for other solid-state units. Much of the service work can be done with an oscilloscope, generator, VOM, and common handtools. However, the equipment capabilities are usually much greater for digital work. Also, there are certain items of test equipment made specifically for digital use. These special test instruments together with the special capability requirements of common test equipment are discussed in the following paragraphs.

4–1.1. Oscilloscopes for Digital Work

An oscilloscope for digital work should have a bandwidth of at least 50 MHz and preferably 100 MHz. The pulse widths or pulse durations found in most digital equipment are on the order of a few microseconds, often only a few nanoseconds wide.

A triggered horizontal sweep is an essential feature. Preferably, the delay introduced by the trigger should be very short. Often, the pulse to be monitored occurs shortly after an available trigger. In other cases, the horizontal sweep must be triggered by the pulse to be monitored. In some oscilloscopes, a delay is introduced between the input and the vertical deflection circuits. This permits the horizontal sweep to be triggered before the vertical signal is applied, thus assuring that the complete pulse will be displayed.

Dual-trace horizontal sweeps are also essential. Most digital troubleshooting is based on the monitoring of two time-related pulses (say, an input pulse and an output pulse or a clock pulse and a readout pulse). With dual trace, the two pulses can be observed simultaneously, one above the other or superimposed if convenient. In other test configurations, a clock pulse is used to trigger both horizontal sweeps. This allows for a three-way time relationship measurement (clock pulse and two circuit pulses, such as one input and one output). A few oscilloscopes have multiple-trace capabilities. However, this is not standard. Usually, such an oscilloscope is provided with plug-in options that increase the number of horizontal sweeps.

The sensitivity of both the vertical and horizontal channels should be such that full-scale deflection can be obtained (without overdriving or distortion) with less than a 1-V signal applied. Typically, the signal pulses used in digital work are on the order of 5 V, but often they are less than 1 V.

In some cases, storage oscilloscopes (for display of transient pulses) and sampling oscilloscopes (for display and measurement of very short pulses) may be required. However, most digital service work can be accomplished with an oscilloscope having the bandwidth and sweep capabilities just described.

4–1.2. Voltmeter for Digital Work

The voltmeters for digital work should have essentially the same characteristics as for other solid-state troubleshooting. For example, a very high input impedance is not critical (20,000 Ω/ V is usually sufficient).

Typically, the circuit operating voltages are 12 V, while the logic voltages (pulse levels) are 5 V, or less. Therefore, the voltmeter should have *good resolution on the low-voltage scales.* For example, a typical logic level might be where $0 = 0$ V and $1 = 3$ V. This means that an input of 3 V or greater to an OR gate will produce a 1 (or true) condition, while an input of something less than 3 V, say 2 V, will produce a 0 (false) condition. As is typical for digital logic specifications, the region between 2 V and 3 V is not spelled out. If the voltmeter is not capable of reading out clearly between 2 and 3 V on some scale, a false conclusion can easily be reached. If the OR gate in question is tested by applying a supposed 3 V, which was actually 2.7 V, the OR gate might or might not operate to indicate the desired 1 or true output.

Generally, d-c voltage accuracy should be ± 2 per cent or better, with ± 3 to ± 5 per cent accuracy for the a-c scales. The a-c scales will probably be used only in checking power supply functions in strictly digital equipment, since all signals are in the form of pulses (requiring an oscilloscope for display and measurement).

The ohmmeter portion of the instrument should have the usual high-resistance ranges. Many of the troubles in the most sophisticated and complex computers boil down to such common problems as cold solder joints, breaks in printed circuit wiring that result in high resistance, as well as shorts or partial shorts between wiring (producing an undesired high-resistance condition). The internal battery voltage of the ohmmeter should not exceed any of the voltages used in the circuits being checked.

4–1.3. Pulse Generators for Digital Work

Ideally, a pulse generator should be capable of duplicating any pulse present in the circuits being tested. Thus, the pulse generator output should be continuously variable (or at least adjustable by steps) in amplitude, pulse duration (or width), and frequency (or repetition rate) over the same range as the circuit pulses. This is not always possible in every case. However, with modern laboratory pulse generators, it is generally practical for most digital equipment.

Typically, the pulses are ± 5 V or less in amplitude but could be 10 V in rare cases. The pulses are rarely longer than 1 sec or shorter than 10 nsec, although there are exceptions here too. Repetition rates are generally less than 100 kHz but could run up to 10 MHz.

Some pulse generators have special features, such as two output pulses with variable delay between the pulses or an output pulse that can be triggered from an external source. However, most routine digital troubleshooting can be performed with standard pulse generators.

4–1.4. Logic Probe

Hewlett-Packard has developed a *logic probe* for digital troubleshooting. This probe, shown in Fig. 4–1, can detect and indicate logic levels or states (0 or 1) in digital circuits. The probe can also detect the presence (or absence) and the polarity of any single pulse 30 nsec or greater in duration. The probe receives power (5 V) from an external source (or the equipment under test) through the cable. The input impedance is 10 K nominal which makes the probe compatible with most TTL (transistor-transistor-logic) and DTL (diode-transistor-logic) digital systems now in common use. There are no operating or adjustment controls. The only indicator is a lamp which appears as a band of light near the probe tip. The probe has a preset threshold of 1.4 V.

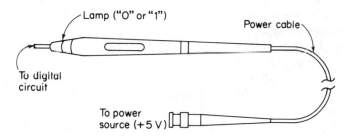

Figure 4–1. Hewlett-Packard logic probe.

When the probe is touched to a high level or is open circuited, a band of light appears around the probe tip. When the probe is touched to a low level, the light goes out. Single pulses of about 30 nsec or wider are stretched to give a light indication time of 0.1 sec. The light flashes on or blinks off, depending on the pulse polarity. When the probe is connected to a pulse train, partial illumination (partial brillance) is displayed by the probe lamp. Pulse trains up to about 1 MHz produce partial brilliance. Pulse trains from 1 to 20 MHz produce either partial brilliance or momentary extinction, depending upon the duty cycle of the pulse. The probe response to different inputs is shown in Fig. 4–2.

When the logic probe is used to detect logic levels, the indicator lamp will be on when the input is high and will be off when the input is low, giving an

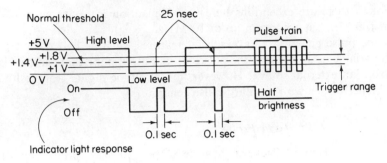

Figure 4–2. Logic-probe response.

indication of a logical 1 or 0, respectively. With power applied and no connection to a circuit, the probe lamp will normally be on.

As shown in Fig. 4–2, the logic probe is ideal for detecting pulses of short duration and low repetition rates that would normally be very difficult to observe on an oscilloscope. Positive pulses 30 nsec or greater in width are stretched and cause the indicator to flash on for 0.1 sec. Negative pulses similarly cause the indicator lamp to be momentarily extinguished. High-frequency pulse trains, too fast for the eye to follow, are indicated by partial illumination. It can be seen that the minimum *on* time of the indicator lamp for positive pulses (or *off* time for negative pulses) is 0.1 sec. The maximum time depends on input pulse width.

Figure 4–3 is the block diagram for the logic probe. Note that both the input circuit and the power supply are protected from overvoltage. Normally, the voltages involved in logic circuits are low (typically, 5 V). However, the probe tip can be touched to a power supply circuit, a power bus for Nixie indicator lamps, or some similar circuit operating at high voltages. Input overload protection is provided from 50 to 200 V for continuous d-c overloads and for 120 V a-c overloads.

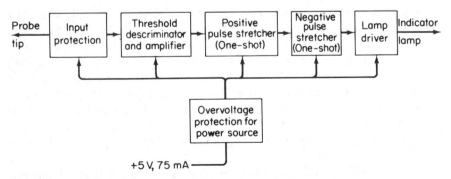

Figure 4–3. Logic-probe block diagram

The threshold discriminator and input amplifier circuit sets the threshold at about 1.4 V. The input amplifier is followed by two pulse stretchers, one of which triggers automatically on each incoming pulse, depending on the pulse polarity. Each stretcher consists of a monostable multivibrator (MV) or one-shot (OS), formed by cross-connecting two gates. When one OS is stretching, the other is acting as an inverting amplifier. The output of the second stretcher drives the indicator lamp through a transistor switch.

The logic probe can be used with several logic circuit analysis techniques. The two most helpful techniques are *pulse-train analysis* and *real-time analysis*. Both techniques are discussed more fully in Sec. 4–2 and 4–3. However, the basis of pulse-train analysis using the logic probe is to run the circuit under test at its normal clock rate, while checking for key logic pulses such as reset, start, transfer, and clock. Questions such as "Is a particular decade counting?" are easily resolved by noting if the probe indicator lamp is partially lit (which occurs only when fast repetition pulse trains are monitored).

With real-time analysis, the normal fast clock signal is replaced by a very slow clock signal from a pulse generator or by manual pulse triggering. The logic changes in the circuit under test will occur at a rate sufficiently slow so that individual level changes and proper pulse occurrences can be observed on a real-time basis.

The logic probe can be tested at any time by touching the tip to a variable d-c source. When the source is above the threshold level, the indicator lamp should be on. When the source is dropped below the threshold level, the indicator lamp should go out.

4–1.5. Logic Clip

Hewlett-Packard has developed a logic clip for digital troubleshooting, where the digital elements are contained in IC form. Although the logic clip will perform all of its functions without additional equipment, it is particularly effective when used with the logic probe (Sec. 4–1.4).

The logic clip, shown in Fig. 4–4, clips onto TTL or DTL ICs (dual in line) and instantly displays the logic states of all 14 or 16 pins. Each of the clip's 16 light-emitting diodes (LED) independently follows level changes at its associated input. A lighted diode corresponds to a high, or 1, logic state.

The logic clip's real value is in its ease of use. The clip has no controls to be set, needs no power connections, and requires practically no explanation as to how it is used. Since the clip has its own gating logic for locating the ground and 5-V_{CC} pins, it works equally well upside down or right-side up. Buffered inputs ensure that the circuit under test will not be loaded down. Simply clipping the unit onto a TTL or DTL dual in-line package of any type makes all logic states visible at a glance.

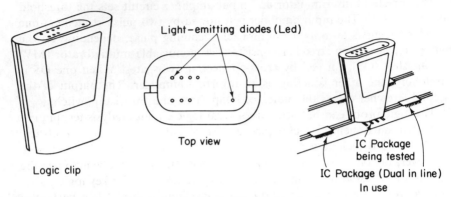

Light-emitting diodes (Led)

Top view

Logic clip

IC Package being tested

IC Package (Dual in line) In use

Figure 4–4. Hewlett-Packard logic clip.

The logic clip is much easier to use than either an oscilloscope or a volt-meter when the troubleshooter is interested in whether a lead is in the *high or low state* (1 or 0), rather than the lead's *actual voltage*. The clip, in effect, is 16 binary voltmeters, and the user does not have to shift his eyes away from his circuit to make the readings. The fact that lighted diodes corresponds to high logic states greatly simplifies the troubleshooting procedure. The user is free to concentrate his attention on his circuits rather than on measure-ment techniques.

When the clip can be used on a real-time basis (when the clock is slowed to about 1 Hz or is manually triggered) timing relationships become especially apparent. The malfunction of gates, FFs, counters, and adders then becomes readily visible as all of the inputs and output of an IC are seen in perspective.

When pulses are involved, the logic clip is best used with the logic probe. Timing pulses can be observed on the probe, while the associated logic-state changes can be observed on the clip.

Figure 4–5 is the block diagram of the logic clip. As shown, each pin of the logic clip is internally connected to a decision gate network, a threshold detector, and a driver amplifier connected to an LED. Figure 4–6 shows the decision sequence of the decision gate network. In brief, the decision gate networks do the following:

1. Find the IC Vcc pin (power voltage) and connect it to the clip power voltage bus. This also activates an LED.
2. Find all logic high pins and activate corresponding LEDs.
3. Find all open circuits and activate corresponding LEDs.
4. Find the IC ground pin, connecting it to the clip ground bus, and blank the corresponding LED.

The threshold detector measures the input voltage. If the voltage is not over

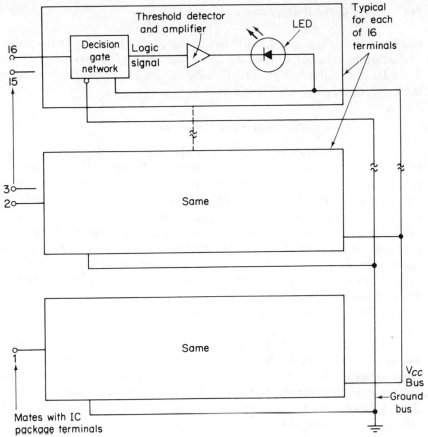

Figure 4–5. Logic-clip block diagram.

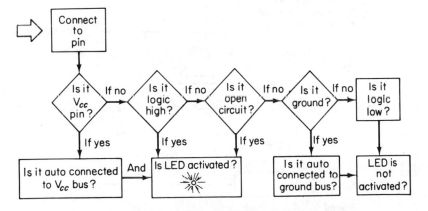

Figure 4–6. Logic-clip decision sequence.

163

the threshold voltage, the LED is not activated. An amplifier at the output
of the threshold detector drives the LED. The LED will indicate high (glow)
if the IC pin is above 2 V and will indicate low (no glow) if pin is below
0.8 V. (If IC pin is open, LED will show high.)

4–1.6. Capacitor Pulse Generator

It is often convenient to trigger logic circuits manually during
troubleshooting. That is, the pulse trains normally present on a particular
circuit line or at a particular input are removed and are replaced by pulses
manually injected one at a time. The obvious procedure is to momentarily
connect the line or input to a bus of appropriate voltage and polarity. This is
not recommended for two reasons. First, some circuits can be damaged by
prolonged application of the voltage. In other circuits, no damage will occur,
but the results will be inconclusive. For example, many faulty circuits will
operate normally when a bus voltage is applied (even momentarily) but will
not respond to a pulse.

A better method is to use a capacitor as a *pulse generator*. The technique is
shown in Fig. 4–7. Simply charge the capacitor by connecting it between
ground and a logic bus (typically 5 V). Then connect the charged capacitor
between ground and the input to be tested. The capacitor will discharge,
creating an input pulse. Make sure to charge the capacitor to the correct
voltage level and polarity. Generally, the capacitor value is not critical. Try
a 0.1-μF capacitor as a starting value. Often, it is convenient to connect one
lead of the capacitor to a ground clip, with the other lead connected to a test

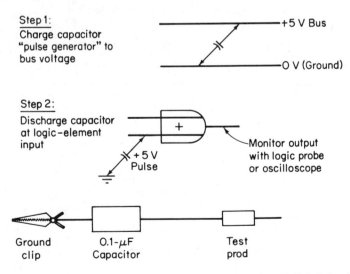

Figure 4–7. Capacitor pulse generator for real-time testing of digital circuits

prod. The capacitor can be clipped to ground, and then the prod tip can be moved from bus to input, as needed.

4–2. BASIC DIGITAL MEASUREMENTS

No matter what equipment is involved or what digital trouble-shooting technique is used, you must be able to work with pulses. That is, you must be able to monitor pulses on an oscilloscope, measure their amplitude, duration or width, and frequency or repetition rate. You must also be able to measure delay between pulses and to check operation of basic digital circuits or building blocks, such as OR gates, AND gates, FFs, delays, and the like.

These basic digital measuring techniques together with descriptions of the test equipment involved are discussed in the following paragraphs.

4–2.1. Basic Pulse Measurement Techniques

The following measurement techniques apply to all types of pulses including squarewaves used in digital circuits.

4-2.1.1. Pulse definitions

The following terms are commonly used in describing pulses found in digital circuits. The terms are illustrated in Fig. 4–8. The input pulse represents an ideal input waveform for comparison purposes. The other waveforms in Fig. 4–8 represent the shape of pulses that may appear in digital circuits, particularly after they have passed through many gates, delays, and so on. The terms are defined as follows.

Rise time, T_R: the time interval during which the amplitude of the output voltage changes from 10 per cent to 90 per cent of the rising portion of the pulse.

Fall time, T_F: the time interval during which the amplitude of the output voltage changes from 90 per cent to 10 per cent of the falling portion of the waveform.

Time delay, T_D: the time interval between the beginning of the input pulse (time zero) and the time when the rising portion of the ouput pulse attains an arbitrary amplitude of 10 per cent above the baseline.

Storage time, T_S: the time interval between the end of the input pulse (trailing edge) and the time when the falling portion of the output pulse drops to an arbitrary amplitude of 90 per cent from the baseline.

Pulse width (or pulse duration), T_W: the time duration of the pulse measured between two 50 per cent amplitude levels of the rising and falling portions of the waveform.

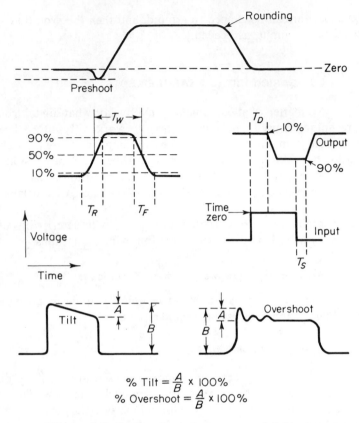

Figure 4–8. Basic pulse and squarewave definitions.

$$\% \text{ Tilt} = \frac{A}{B} \times 100\%$$
$$\% \text{ Overshoot} = \frac{A}{B} \times 100\%$$

Tilt: a measure of the tilt of the full-amplitude, flat-top portion of a pulse. The tilt measurement is usually expressed as a percentage of the amplitude of the rising portion of the pulse.

Overshoot or preshoot: a measure of the overshoot or preshoot occurring generally above or below the 100 per cent amplitude level. These measurements are also expressed as a percentage of the pulse rise.

These definitions are for guide purposes only. When pulses are very irregular (such as excessive tilt, overshoot, etc.), the definitions may become ambiguous.

4-2.1.2. Rule of Thumb for Rise-time Measurements

Since rise-time measurements are of special importance in any pulse circuit testing, the relationship between the oscilloscope rise time and the rise times of the digital circuit components under test must be taken into account. Obviously, the accuracy of rise-time measurements can be no greater than the rise time of the oscilloscope. Also, if the device is tested by means of an external pulse from a pulse generator, the rise time of the pulse generator must also be taken into account.

For example, if an oscilloscope with a 20-nsec rise time is used to measure the rise time of a 15-nsec digital circuit component, the measurement will be hopelessly inaccurate. If a 20-nsec pulse generator and a 15-nsec oscilloscope is used to measure the rise time of a component, the fastest rise time for accurate measurement will be something greater than 20 nsec. Two basic rules of thumb can be applied to rise time measurements.

The first method is known as the *root of the sum of the squares*. It involves finding the squares of all the rise times associated with the test, adding these squares together, and then finding the square root of the sum. For example, using the 20-nsec pulse generator and the 15-nsec oscilloscope, the calculation is

$$20 \times 20 = 400, \quad 15 \times 15 = 225, \quad 400 + 225 = 625,$$
$$\sqrt{625} = 25 \text{ nsec}$$

This means that the fastest possible rise time capable of measurement is 25 nsec.

One major drawback to this rule is that the coaxial or shielded cables required to interconnect the test equipment are subject to *skin effect*. As frequency increases, the signals tend to travel on the outside or skin of the conductor. This decreases conductor area and increases resistance. In turn, this increases cable loss. The losses of cables do not add properly to apply the root-sum-squares method, except as an approximation.

The second rule or method states that if the equipment or signal being measured has a rise time ten times slower than the test equipment, the error is 1 per cent. This amount is small and can be considered as negligible. If the equipment being measured has a rise time three times slower than the test equipment, the error is slightly less than 6 per cent. By keeping these relationships in mind, the results can be interpreted intelligently.

4-2.1.3. Measuring Pulse Amplitude

Pulse amplitudes are measured on an oscilloscope, preferably a laboratory-type oscilloscope where the vertical scale is calibrated directly in a specific deflection factor (such as volts per centimeter). Such oscilloscopes usually have a step attenuator (for the vertical amplifier) where each step is related to a specific deflection factor. Typically, the pulses used in digital circuits are 5 V or less. Often, the pulse amplitude is critical to the operation of the equipment. For example, an AND gate may require two pulses of

3 V to produce an output. If the two pulses are slightly less than 3 V, the AND gate may act as if there is no input or that both inputs are false. Thus, when troubleshooting any digital circuit, always check the actual pulse amplitude against that shown in the service literature. Generally, digital circuits are designed with considerable margin in pulse amplitude, but equipment aging or some malfunction can reduce the safety margin.

The basic test connection and corresponding oscilloscope display for pulse amplitude measurement are shown in Fig. 4–9. The basic procedure is as follows:

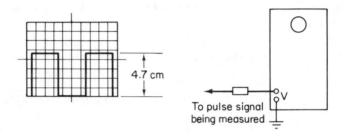

Figure 4–9. Measuring pulse amplitude.

1. Connect the equipment as shown in Fig. 4–9. Place the oscilloscope in operation.

2. Set the vertical step attenuator to a deflection factor that will allow the expected signal to be displayed without overdriving the vertical amplifier. Typically, the vertical deflection factor should be 10 V or less full scale.

3. Set the input selector to measure a-c. Pulses can also be measured with the oscilloscope set to d-c. However the trace will be moved up or down on the screen, depending on the polarity of the d-c voltage present with the pulse.

4. Connect the probe to the pulse signal being measured. Set the oscilloscope controls to display at least one and preferably two or more pulses on the screen. In most cases, this involves switching on the internal recurrent sweep such that the sweep is driven by the pulse train being measured. If the pulse being measured is of the one-shot nature where there is a relatively long time between pulses, it may be necessary to use an oscilloscope where there is a delay introduced between the beginning of the sweep and the vertical display. However, this is rare in most digital circuit troubleshooting.

5. Adjust the controls to spread the pattern over as much of the screen as desired. Adjust the vertical position control so the downward excursion of the waveform coincides with one of the screen lines below the centerline, as shown in Fig. 4–9.

6. Adjust the horizontal position control so that one of the upper peaks of

the pulse (which presumably has a flat top) lies near the vertical centerline, as shown in Fig. 4–9.

7. Measure the peak-to-peak vertical deflection in centimeters. Multiply the distance measured by the vertical attenuator switch setting. Also include the attenuation factor (if any) of the probe. Generally, digital pulse circuits are measured with a low-capacitance probe having no attenuation or where the attenuation is included as part of the vertical attenuator switch setting.

As an example, assume that a vertical deflection of 4.7 cm (Fig. 4–9) is measured, using a 1X probe (no attenuation), and a vertical deflection factor (step attenuator setting) of 1 V/cm. Then the peak-to-peak pulse amplitude is 4.7 × 1 × 1=4.7 V.

4-2.1.4. Measuring Pulse Width or Duration

Pulse width or pulse duration is measured on an oscilloscope, preferably a laboratory type, where the horizontal scale is calibrated directly in relation to time. The horizontal sweep circuit of a laboratory oscilloscope is usually provided with a selector control that is direct reading in relation to time. That is, each horizontal division on the screen has a definite relation to time at a given position of the horizontal sweep rate switch (such as 1 μsec/cm). Typically, the pulses used in digital circuits are on the order of nanoseconds or microseconds, but can be as long as a few milliseconds.

The basic test connection and corresponding oscilloscope display for pulse width measurement is shown in Fig. 4–10. The basic procedure is as follows:

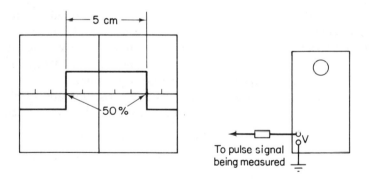

Figure 4–10. Measuring pulse width.

1. Connect the equipment as shown in Fig. 4–10. Place the oscilloscope in operation.

2. Set the vertical step attenuator to a deflection factor which will allow the expected signal to be displayed without overdriving the vertical amplifier. Set the input selector to measure a-c.

3. Connect the probe to the pulse signal being measured. Switch on the oscilloscope internal recurrent sweep. Set the horizontal sweep control to the fastest sweep rate that will display a convenient number of divisions between the measurement points (Fig. 4–10) On most oscilloscopes, it is recommended that the extreme sides of the screen not be used for width or time-duration measurements. There may be some nonlinearity at the beginning and end of the sweep.

4. Adjust the vertical position control to move the measurement points to the horizontal centerline. Adjust the horizontal position control to align one edge (usually the left-hand edge) of the pulse with one of the vertical lines.

5. Measure the horizontal distance between the measurement points on the left- and right-hand edges of the pulse. If width is the only concern, use the 50 per cent points shown in Fig. 4–8. If rise time is to be measured, use the 10 and 90 per cent points. Better resolution is obtained if the 10 and 90 per cent points are used.

6. Multiply the distance measured by the setting of the horizontal sweep control. Also include the sweep magnification factor (if any). Due to the very short pulse durations, it is often necessary to use sweep magnification.

As an example, assume that the horizontal sweep rate is 0.1 msec/cm, the sweep magnification is 100, and a horizontal distance of 5 cm is measured (between the 50 per cent points). Then the pulse width is $5 \times 0.1/100 = 0.005$ msec or 5 μsec.

4–2.1.5. Measuring Pulse Frequency or Repetition Rate

Pulse frequency or repetition rate is measured on an oscilloscope, preferably a laboratory type, where the horizontal scale is calibrated directly in relation to time. Pulse frequency is found by measuring the time duration of one complete pulse cycle (not pulse width) and then dividing the time into 1 (since frequency is the reciprocal of time).

The basic test connection and corresponding oscilloscope display for pulse-frequency measurement is shown in Fig. 4–11. The basic procedure is as follows:

1. Connect the equipment as shown in Fig. 4–11. Place the oscilloscope in operation.

2. Set the vertical step attenuator to a deflection factor which will allow the expected signal to be displayed without overdriving the vertical amplifier. Set the input selector to measure ac.

3. Connect the probe to the pulse signal being measured. Switch on the oscilloscope internal recurrent sweep. Set the horizontal sweep control so that one complete pulse cycle is displayed. Avoid using the extreme sides of the screen. If necessary, display more than one complete cycle.

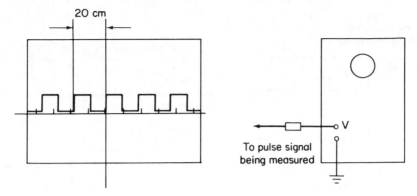

Figure 4–11. Measuring pulse frequency or repetition rate.

4. Adjust the vertical position control to move the measurement points to the horizontal centerline. Adjust the horizontal position control to align one measurement point on the pulse with one of the vertical lines. In Fig. 4–11, the start of the left hand edge of the pulse is used as the measurement point.

5. Measure the horizontal distance between the measurement points. Multiply the distance measured by the setting of the horizontal sweep control. Also include the sweep magnification factor (if any). Divide the time measured into 1 to find frequency.

As an example, assume that the horizontal sweep rate is 0.1 msec/cm, the sweep magnification is 100, and a horizontal distance of 20 cm is measured between the beginning and end of a complete cycle. Then the frequency is $20 \times 0.1/100 = 0.020$ msec or 20 μsec; $1/20\ \mu$sec $= 50$ kHz.

4–2.1.6. Measuring Pulse Times with External Timing Pulses

It is possible to use an external *time-mark generator* to calibrate the horizontal scale of an oscilloscope for pulse-time measurement. Time-mark generators are used for many reasons, but their main advantages for digital applications are greater accuracy and resolution. For example, a typical crystal-controlled, time-mark generator will produce timing signals at intervals of 10 and 1 μsec, as well as at 100 nsecs.

A time-mark generator produces a pulse-type timing wave which is a series of sharp spikes spaced at precise time intervals. These pulses are applied to the oscilloscope vertical input and appear as a wavetrain similar to that shown in Fig. 4–12. The oscilloscope horizontal gain and positioning controls are adjusted to align the timing spikes with screen lines, until the screen divisions equal the timing pulses. The accuracy of the oscilloscope timing circuits is then of no concern, since the horizontal channel is calibrated against the external time-mark generator.

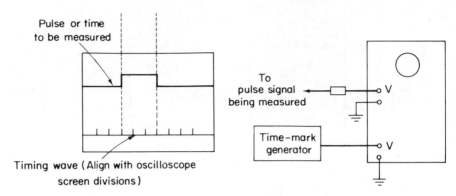

Figure 4–12. Measuring pulse time with external timing pulses.

The timing pulses can be removed, and the pulse signal to be measured can be applied to the vertical input, provided that the *horizontal gain and positioning controls are not touched.*

Duration or time is read from the calibrated screen divisions in the normal manner. If the oscilloscope has a dual-trace feature, the time-mark generator can be connected to one input; the other input receives the pulse signal to be measured. The two traces can be superimposed or aligned, whichever is convenient.

4–2.1.7. Measuring Pulse Delay

The time interval or delay between two pulses (say, an input pulse and an output pulse) introduced by a delay line, FF, or other digital circuit can be measured most conveniently on an oscilloscope with a dual trace. It is possible to measure delay on a single-trace oscilloscope, but it is quite difficult. If the delay is exceptionally short, the screen divisions can be calibrated with an external time-mark generator. If the oscilloscope is a dual-trace instrument, the screen divisions must be calibrated against the time-mark generator as described in Sec. 4–2.1.6. Then the time-mark generator can be removed, and the input and output pulses can be applied to the two vertical channels.

The basic test connection and corresponding oscilloscope display for pulse-delay measurement is shown in Fig. 4–13. The basic procedure is as follows:

1. Connect the equipment as shown in Fig. 4–13. Place the oscilloscope in operation. Switch on the internal recurrent sweep.
2. Set the sweep frequency and sync controls for a single, stationary input pulse and output pulse as shown in Fig. 4–13. Set the horizontal and vertical

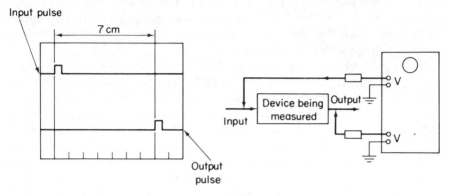

Figure 4–13. Measuring pulse delay.

gain controls for the desired pulse pattern, width, and height. Avoid the extreme limits of the screen.

3. Count the screen divisions between the *same point* on both the input and output pulses. This is the delay interval.

As an example, assume that the horizontal sweep rate is 100 nsec/cm, there arc seven screen divisions between the input and output pulses, and each horizontal screen division is 1 cm. Then the delay is $7 \times 100 = 700$ nsec.

4–2.2. Testing Digital Circuits

The following paragraphs describe procedures for test or checkout of basic digital circuits (gates, delays, FFs, etc.). Such circuits can be tested on a real-time basis or with pulse trains. Both test methods are described.

An oscilloscope is the best tool for pulse-train testing. In its simplest form, pulse-train testing consists of measuring the pulses at the input and output of a digital circuit. The pulses can be those normally present in the circuit, or they can be introduced from an external pulse generator. If the input pulses are normal but the output pulses are absent or abnormal, the fault is most likely to be in the circuit being tested. An exception to this is where the following circuit is defective (say, a short circuit) and makes the output of the circuit under test *appear* to be absent or abnormal. Either way, the trouble is localized to a specific point in the overall circuit.

If the input pulses to the circuit under test are absent or abnormal, the trouble is most likely to be in circuits ahead of the point being tested. Thus, when input pulses are found to be abnormal at a certain circuit, this serves as a good starting point for troubleshooting.

The Hewlett-Packard logic probe (Sec. 4–1) can also be used for pulse-train analysis. However, the logic probe will indicate only that a *pulse train exists at a given point*. The logic probe cannot tell the pulse width, frequency, amplitude, or whether the pulse is in step with other pulses. Only an oscilloscope will show this information.

The logic probe and logic clip are ideal tools for *real-time analysis*. Such testing can take several forms. The clock frequency can be slowed to some very low rate (typically 1 Hz), so that individual input and output pulses can be monitored individually (on the probe, clip, or oscilloscope). The clock rate is slowed by substituting an external pulse generator for the normal (internal) clock signal. If this is not convenient, it is possible to inject pulses from the generator directly at the inputs of a given circuit, or pulses can be introduced with a charged capacitor (Sec. 4–1.6).

Keep in mind that digital troubleshooting usually starts with a test of several circuits simultaneously. There may be several hundred (or thousand) circuits in a digital device. It could take months to check each circuit on an individual basis. However, there are tests that check whole groups of circuits simultaneously. The self-check or operational-check procedures found in digital equipment service literature are based on this technique. Self-check techniques are described in Sec. 4.3.

A classic example of testing several digital circuits simultaneously is when four FFs are connected as a decade counter. If there is an output pulse train at the fourth FF and this pulse train shows one pulse for every ten input pulses at the first FF, the entire counter circuit can be considered as good.

Also keep in mind that many digital circuits are often combined in one IC package. Even if it is possible to test the individual circuits, they cannot be replaced. The complete IC function must be checked as a package by monitoring inputs and outputs at the IC terminals.

Before going into detailed test procedures, the *marginal test technique* should be considered. The power supplies for most digital circuits are adjustable, usually over a ± 10 per cent range. When digital equipment failure is intermittent or the failure is not consistant, it is sometimes helpful to test the circuits with the power supply output voltage set to extremes (both high

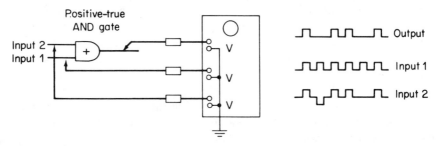

Figure 4–14. Testing AND gate circuits.

and low). This will show marginal failures in some cases. The technique is not recommended by all manufacturers since the results are not certain. However, it is usually safe to apply the technique to any equipment on a temporary basis. Of course, the equipment service literature should be consulted to make sure.

The following procedures are based on the assumption that the corresponding digital circuit can be tested on an individual basis.

4–2.2.1. Testing AND Gate Circuits

Figure 4–14 shows the basic connections for testing an AND circuit. Ideally, both inputs and the output should be monitored simultaneously (with a multiple-trace oscilloscope), since an AND circuit produces an output only when two inputs are present. If pulse trains are monitored on an oscilloscope, check that an output pulse is produced each time there are two input pulses of appropriate amplitude and polarity. If this is not the case, check carefully that both input pulses arrive at the same time (not delayed from each other).

If the oscilloscope has a dual-trace feature, first check for a series of simultaneous pulses at both inputs. Then move one oscilloscope probe to the output and check for an output pulse. Note that in Fig. 4–14 the input pulse trains are not identical which is often the case. Input 2 has fewer pulses for a given time interval than input 1. Also, input 2 has one negative pulse simultaneously with one positive pulse at input 1. This should not produce an output pulse. Often, one input will be a long pulse (long in relation to the pulses at the other input). In other cases, one input will be a fixed d-c voltage applied by operation of a switch.

If the AND circuit is tested with pulse trains and a logic probe, simply check for a pulse train at both inputs and at the output. If the circuit is checked on a real-time basis, inject simultaneous pulses at both inputs (with a capacitor generator or whatever is convenient) and monitor the output with a logic probe or oscilloscope.

The complete truth table of the circuit can be checked using the basic connections of Fig. 4–14. For example, an output pulse train (or single pulse in the case of real-time testing) should not appear if only one input of appropriate polarity is present.

4–2.2.2. Testing OR Gate Circuits

Figure 4–15 shows the basic connections for testing an OR circuit. Ideally, all inputs and the output should be monitored simultaneously. An OR gate produces an output when any inputs are present. If pulse trains are monitored on an oscilloscope, check that an output pulse is

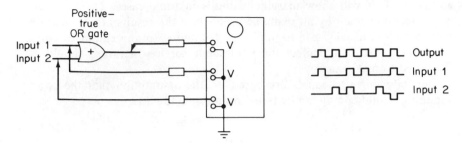

Figure 4–15. Testing OR gate circuits.

produced for each input pulse of appropriate amplitude and polarity.

If the oscilloscope has a dual-trace feature, monitor the output with one trace; then monitor each of the inputs in turn with the other trace. Note that in Fig. 4–15 the input pulse trains are not identical. That is, the input pulses do not necessarily coincide. However, there is an output pulse when either input has a pulse and when both inputs have a pulse. This last condition marks the difference between an OR gate and an EXCLUSIVE OR gate, discussed in Sec. 4-2.2.3.

If the OR circuit is tested with pulse trains and a logic probe, check that there is a pulse train at the output whenever there is a pulse at any input. If the circuit is checked on a real-time basis, monitor the output and inject a pulse at each input, in turn. The complete truth table of the circuit can be checked using the basic connections of Fig. 4–15.

4–2.2.3. Testing EXCLUSIVE OR Gate Circuits

Figure 4–16 shows the basic connections for testing an EXCLUSIVE OR circuit. Ideally, all inputs and the output should be monitored simultaneously. An EXCLUSIVE OR gate produces an output when any one input is present but not when both inputs are present. If pulse trains are

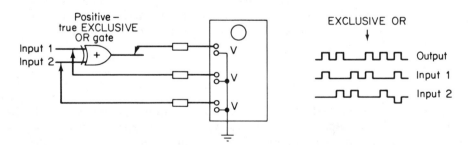

Figure 4–16. Testing EXCLUSIVE OR gate circuits.

monitored on an oscilloscope, check that an output pulse is produced for each input pulse that does not coincide with another input pulse.

If the oscilloscope has a dual-trace feature, first monitor two inputs with one trace. Note if there is any condition in the pulse train where only one input pulse occurs at a given time. Then monitor the output and check for an output pulse that corresponds to the single input pulse.

If the EXCLUSIVE OR circuit must be checked with a logic probe, it will be necessary to do so on a real-time basis. The logic probe can indicate the presence of pulse trains but not coincidence of pulses in pulse trains. Monitor the output and inject a pulse at each input, in turn. An output pulse should be produced for each input pulse. Then apply simultaneous pulses to both inputs. There should be no output. The complete truth table of the circuit can be checked using the basic connections of Fig. 4–16.

4–2.2.4. Testing NAND and NOR Gate Circuits

Figures 4–17 and 4–18 show the basic connections for testing NAND and NOR circuits, respectively. Note that the connections are the same for corresponding AND and OR circuits as are the test procedures. However, the output pulse will be inverted. That is, a NAND circuit produces an output under the same logic conditions as an AND circuit, but the output

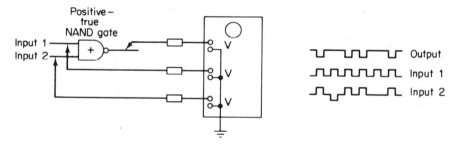

Figure 4–17. Testing NAND gate circuits.

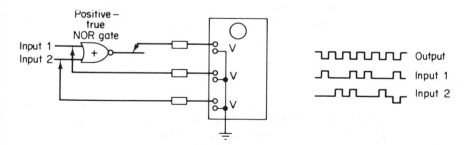

Figure 4–18. Testing NOR gate circuits.

will be inverted. For example, if the logic is positive true and a positive pulse is present at both inputs of a NAND gate simultaneously, there will be an output pulse, and the pulse will be negative (or false).

4–2.2.5. Testing ENCODE Gate Circuits

Figure 4–19 shows the basic connections for testing an EN-CODE circuit. Ideally, all outputs and the input should be monitored simultaneously. An ENCODE gate produces multiple outputs simultaneously which correspond in polarity or logic state to the input.

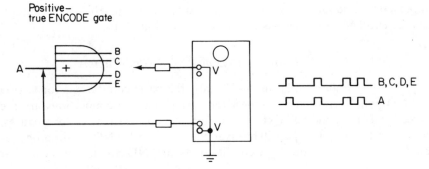

Figure 4–19. Testing ENCODE gate circuits.

If the pulse trains are monitored on an oscilloscope with a dual-trace feature, monitor the input with one trace; then monitor each of the outputs, in turn, with the other trace. If the ENCODE circuit is tested with pulse trains and a logic probe, check that there is a pulse train at each output whenever there is a pulse train at the input. If the circuit is checked on a real-time basis, inject a pulse at the input and monitor each output.

4–2.2.6. Testing Amplifiers, Inverters, and Phase splitters

Figure 4–20 shows the basic connections for testing amplifiers, inverters, and phase splitters used in digital pulse circuits. Note that the same connections are used for all three circuits. That is, each circuit is tested by monitoring the input and output with an oscilloscope or logic probe. However, the relationship of output to input is different for each of the three circuits.

With the amplifier, the output pulse amplitude should be greater than the input pulse amplitude. With the inverter, the output pulse may or may not be greater in amplitude than the input pulse, but the polarity will be reversed. In the case of the phase splitter, there will be two output pulses of opposite polarity to each other for each input pulse.

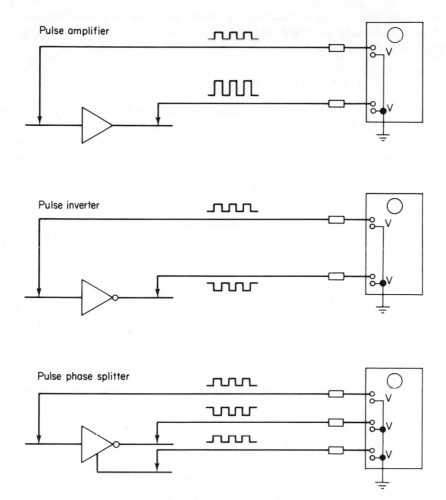

Figure 4–20. Testing amplifiers, inverters, and phase splitters.

Polarity will be easy to determine if the pulses are monitored on an oscilloscope. The ability of the logic probe to show polarity depends on the logic system used. The logic probe responds to systems where the logic levels are between 0 and 5 V. That is, a true or high level is 5 V, while a false or low level is 0 V. The logic probe will blink on for each pulse, if the pulse goes from some level below about 1.4 V to a level above 1.4 V and will blink off when the pulse drops from a high level to some voltage below about 1.4 V. If the probe is used to check an inverter where the input goes from 0 to 5 V and the output goes from 5 to 0 V, the input pulse train will produce a series of "on blinks," while the output pulse train produces "off blinks." The difference between on and off may be difficult to distinguish if the pulse trains are fast. It may be necessary to make such tests on a real-time basis. Also, if the inverter produced an output that went from 0 to − 5 V, the logic probe would not respond properly.

4-2.2.7. Testing FF Circuits

Figure 4–21 shows the basic connections for testing an FF. As discussed in Chapter 3, there are many types of FFs (R-S, J-K, toggle, latching, delay, etc.). Each FF responds differently to a given set of pulse conditions. Thus, each type of FF must be tested in a slightly different way.

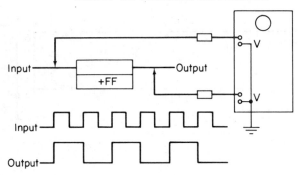

Figure 4–21. Testing FF circuits.

However, all FFs have two states (even though there may be only one output, and it may not be possible to monitor both states directly). Some FFs require only one input to change states. Other FFs require two simultaneous pulses to change states (such as a clock pulse and a reset or set pulse).

The main concern in testing any FF is that the states change when the appropriate input pulse (or pulses) are applied. Thus, at least one input and one state (output) should be monitored simultaneously. For example, an R-S FF (without clock) can be tested by monitoring the pulse train at the set input and set output. If there is a pulse train at the set input and a set output pulse train, it is reasonable to assume that the FF is operating. If the set output remains in one state, either the FF is defective and not being reset or there are no reset pulses. The next step is to monitor the reset input and output. If both the set and reset trains are present but the FF does not change states, the FF is defective.

There are exceptions, of course. For example, if the reset and set pulses arrive simultaneously due to an unwanted delay in other circuits, the FF may remain in one state.

If the same R-S FF requires a clock pulse, first check that both the clock pulse and a set or reset pulse arrive simultaneously. Then check that the FF changes states each time there is such a pulse coincidence. For example, assume that pulse trains are measured on a dual-trace oscilloscope and that a set pulse occurs for every other clock pulse. Next, monitor the clock pulse and the set output, and note that the set output occurs for every two clock pulses. Or, monitor the set input and output, noting that the FF changes states on a pulse-for-pulse basis.

To test an FF with pulse trains and a logic probe, simply check that a pulse train exists at each output, with a pulse train at the inputs. It can then be assumed that the FF is operating properly. To remove all doubt, use the logic probe to test the FF on a real-time basis. Inject a pulse at the set input and monitor the set output. Then inject a reset input and note a change in states.

4–2.2.8. Testing MVs

Figure 4–22 shows the basic connections for testing an MV. As discussed in Chapter 3, there are three basic types of MVs: free-running, one-shot (monostable), and Schmitt trigger (bistable). In the case of the free-running MV, only the output need be monitored since the circuit is self-generating. Both the one-shot (OS) and Schmitt trigger (ST) require that the input and output be monitored.

If a logic probe is used to monitor pulse trains from an MV, the presence of a pulse train at the output indicates that the MV is *probably* working. The OS and ST MVs can be tested on a real-time basis as well by injecting pulses

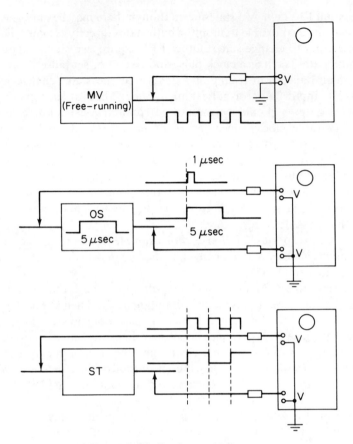

Figure 4–22. Testing multivibrators.

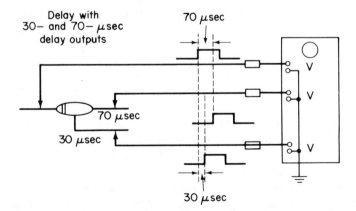

Figure 4–23. Testing delays.

at the input. Keep in mind that the ST requires two input pulses for a complete cycle. That is, two input pulses are required to change the output states and then return the states back to the original condition. More simply, two input pulses are required for one complete output pulse. The OS MV produces one complete output pulse for each input pulse, even though the output pulse may be different in duration from that of the input pulse.

If MV pulse trains are monitored on an oscilloscope, the frequency, pulse duration, and pulse amplitude can be measured as described in Sec. 4–2.1. While these factors may not be critical for all digital circuits, quite often one or more of the factors is important. For example, the logic symbol for an OS MV (when properly drawn) includes the output pulse duration or width.

4–2.2.9. Testing Delays

Figure 4–23 shows the basic connections for testing delay lines or delay elements used in digital circuits. The test procedure is the same as for measurement of delay between pulses described in Sec. 4–2.1.7.

Keep in mind that the delay shown on the logic symbol *usually* refers to the delay between leading edges of the input and output pulses. This can be assumed if the symbol is not further identified. Some manufacturers (such as Hewlett-Packard) identify their delay symbols as to leading or trailing edge, particularly if the symbols are mixed on a single logic diagram.

4–3. DIGITAL EQUIPMENT SERVICE LITERATURE

Unlike some other electronic equipment, it is almost impossible to troubleshoot digital circuits without adequate service literature. There are two basic reasons for this.

Although all digital equipment use the same basic circuits (gates, FFs, delays, etc.), the way in which these circuits are arranged to perform an overall function is unique to each type of digital equipment. Thus, it is absolutely essential that you know exactly what the digital equipment is to do from an operational standpoint. Further, you must know what pulse will appear at what point and under what conditions.

Another reason for good service literature is that, in servicing digital equipment, you are dealing with failure rather than poor performance. For example, a television set may have a weak picture or some distortion in the audio. This may pass unnoticed or be ignored during normal operation. On the other hand, a digital counter will either read out a correct count, fail to read out a correct count, or provide no readout. Anything less than the correct count is not acceptable. You *must know* what is correct in every case.

Knowing the equipment can come only after a thorough study of the service literature. In the case of very complex digital devices (particularly computers), you must also receive factory training before you can hope to do a good troubleshooting job.

All the information found in service literature is of value, but the most important pieces of information for troubleshooting are the logic diagram, the theory of operation, and the self-check procedure. Also, timing diagram can be of considerable help.

The value of the logic diagram and theory of operation for troubleshooting is obvious. To trace pulses throughout the equipment, you must have a logic diagram, preferably supplemented with individual schematics that show the internal circuits of the *blocks* on the overall diagram. The individual schematics are often omitted when the blocks are IC packages. To understand the logic diagram, you must have a corresponding theory of operation (unless you are very familiar with similar equipment, or you just happen to be a genius).

The value of a timing diagram is sometimes less obvious. However, trouble can often be pinpointed immediately by comparison of circuit pulses against those shown in the timing diagram. Also, unusual problems can show up when pulses are compared with the timing diagram. For example, assume that a counter receives input (or count) pulses and reset pulses, both of which originate from a common clock or master oscillator. The reset pulses are supposed to arrive at the counter at some specific time interval thus allowing a given maximum number of clock pulses (say, 100) to be read out. Further, assume that a defect in the circuit has caused the reset pulses to be delayed slightly in relation to the clock pulse so that the reset pulse coincides with the first count. The counter will never be able to reach a second count, and the readout will always remain at zero, even though all of the circuits appear to be good.

The value of a self-check procedure depends on the complexity of the equipment. On simple equipment, an ingenious technician can sometimes devise a check procedure that will eliminate groups of digital circuits from suspicion or pinpoint the problem to a group of circuits. For example, assume that a digital counter circuit is to be checked. The counter gate can be set to open for 1-sec intervals, with a 1-kHz pulse signal introduced at the counter input. If the counter is operating properly, the readout will be 1000. (From a practical standpoint, if the readout is 999 or 1001, the counter may still be functioning properly. Most digital counters have a ± 1 count ambiguity.) The gate time is then shortened or lengthened, and the input pulse frequency is changed (increased or decreased), to check all of the digits in the readout or each digit one at a time, whichever is convenient.

On complex equipment, such as a computer, the self-check must be thought out carefully. Unless you are completely familiar with every logic equation, you will do much better to follow the service literature. Usually, a computer self-check involves inserting a series of equations to be solved and then noting the corresponding readout. On those computers with banks of interpolation lights (lights that show the state of binary FFs, on or off, 1 or 0), equations are inserted and the lights are checked as to their state. On other computers, the readout is printed on a typewriter or tape punch. Either way, the equations and the corresponding readouts are compared line for line, until an abnormal readout is found. The service literature usually contains some means of relating the equation and/or readout to a specific circuit or group of circuits.

Very sophisticated computers often have an automatic self-check provision. When the computer is operating but is at rest (not solving a specific problem), the circuits shift over to a self-check and provide a continuous readout to the operator.

4–3.1. Examples of Service Literature

The following paragraphs show typical examples of information found in digital equipment service literature. This information is extracted from the operating and service manual for a Hewlett-Packard scanner. The information is summarized. Full data appear in the manual. However, the data presented here are sufficient to familiarize the reader with what is available in well-prepared service literature, and how the data can be related to troubleshooting.

The purpose of the scanner is to sample several channels of information and pass this information to a single output on a time-sharing basis (channel after channel). In addition to many other controls and circuits, the scanner has a digital readout consisting of three Nixie tubes. The readout indicators show the particular channel (or *address* as it is called in the manual) being monitored. (For scanning functions, the terms address and channel are essentially synonymous; in computer applications, address is the common term; in analog scanning, channel is generally used.)

4–3.1.1. Logic Diagrams and Supplementary Troubleshooting Data

Figure 4–24 shows portions of the scanner logic diagram, particularly those portions covering the digital readout circuits. Note that each Nixie tube has a separate 4/10-line decoder and a decade counter. These are shown as blocks in Fig. 4–24. since they are IC packages. The internal circuitry of these particular blocks is not shown anywhere in the manual, since the ICs must be replaced as a package. However, the logic diagram is supplemented by two diagrams, Figs. 4–25 and 4–26, which show the IC

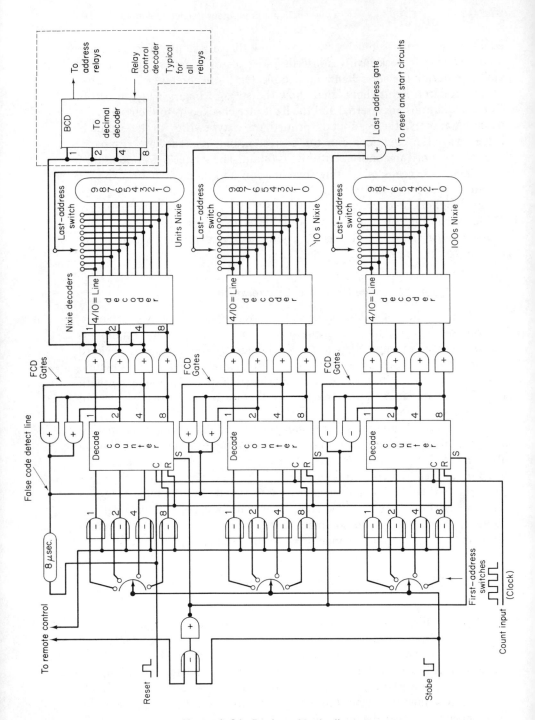

Figure 4-24. Portion of logic diagram.

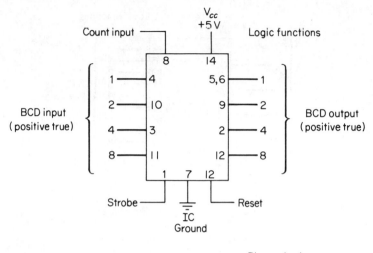

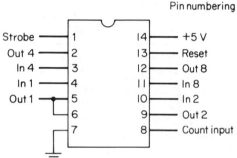

Figure 4–25. Logic functions and pin-numbering diagram of decade counter.

package terminal arrangements and the outputs or logic states to be expected for a given set of input conditions. By using this supplementary data, it is possible to check the IC packages on an individual basis with real-time pulses (using a logic clip or oscilloscope). Or, the IC packages can be checked in operation by monitoring pulse trains at the inputs and outputs.

The following are examples of how the IC packages can be checked on an individual basis, while still installed in the circuit. This is done when a particular IC is suspected of being defective (as a result of self-check and/or logical deductions, described in Sec. 4–3.1.3.).

To check the decade counter of Fig. 4–25 on a real-time basis using a logic clip, proceed as follows:

1. Install the logic clip on the IC to be tested by squeezing the thick end of the clip to spread the contacts and placing the clip on the IC.

Logic functions

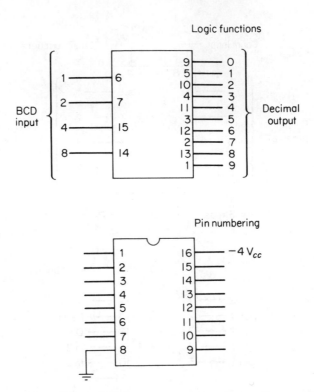

Pin numbering

Figure 4–26. Logic functions and pin-numbering diagram of 4/10 decoder.

2. Reduce the clock frequency to some rate that can be seen easily (about 1 Hz or slower) by substituting a pulse generator input for the clock oscillator. The scanner in question can be slowed down (by a front panel control) so that there are about 10 sec between each channel. Thus, no external pulse is needed to produce a slow clock rate. As an alternative to slowing the clock rate, remove the IC package connections to the input (IC pin 8), strobe (IC pin 1), and reset (IC pin 13). Then inject pulses at these inputs using a capacitor generator or whatever is convenient.

3. Inject a pulse at reset (IC pin 13) and check that all four output pins are false (LEDs on logic clip go out).

4. Inject a series of pulses at the input (IC pin 8) and check that the output pins go true (LEDs go on) in accordance with the logic in Fig. 4–25. For example, with one input pulse, the 1 output goes on. With two input pulses, the 1 goes out and the 2 goes on. With three input pulses, the 2 remains on and the 1 goes back on.

5. Inject a pulse at the strobe (IC pin 1) and check that the BCD output pins assume the same states as the BCD input pins. Note that the strobe function is synonymous with an enable function. That is, the outputs assume the

same states as the inputs whenever the strobe or enable pulse is applied. If all of the inputs are false, inject a pulse at one of the inputs and make sure the coresponding output assumes the same state. For example, if a pulse is applied to the 8 input (IC pin 11) and the strobe (IC pin 1) simultaneously, the 8 output (IC pin 12) should go true.

To check the decade counter of Fig. 4–25 on a pulse-train basis with a dual-trace oscilloscope, proceed as follows:

1. Monitor the input (IC pin 8) and the reset (IC pin 13). If there are no reset pulses, it is safe to proceed with the next step. If there are reset pulses, count the number of input pulses between reset pulses. If the count is ten or more, the decade can reach a full ten count. If there are less than ten input pulses between reset pulses, this sets the maximum count. For example, if there are seven input pulses between reset pulses, the counter can reach a maximum count of seven. Thus, the 8 output (IC pin 12) will not show any pulses (unless the counter is defective).

2. Monitor the strobe (IC pin 1) and each of the BCD inputs in turn. If there is a strobe pulse that coincides with any of the BDC input pulses, check that the corresponding BCD output line also has pulses that coincide with the strobe pulses.

4–3.1.2. Timing Diagrams and Theory of Operation

The test procedures just described establish that the IC decade package responds properly to a set of pulses. However, the tests do not show that the pulses are correct (arrive at the right place at the right time). This can be determined only by reference to the timing diagram, logic diagram, and theory of operation. Figure 4 27 is the timing diagram for that portion

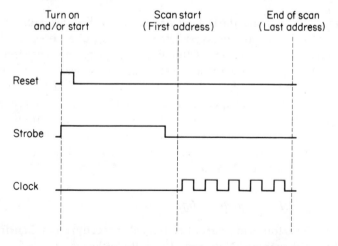

Figure 4–27. Timing diagram for decade-counting function.

of the scanner logic shown in Fig. 4–24. The following paragraphs summarize the theory of operation that applies to the same circuits.

Information to be scanned is applied to the common output through relays. There is one relay for each channel. Each relay is connected to a separate line on the BCD-to-decimal decoder. When an output (pulse or continuous voltage) is present on one of these lines, the corresponding relay is operated, and the information on that channel is passed to the common output. The address being monitored at any given time is indicated by the Nixie readout tubes which are operated by separate decoders.

Several modes of operation are available. In the continuous scan mode, a first and last channel (or first and last address) are selected by front panel switches. The scanner starts by monitoring the first address and then monitors each address in turn until the last address is reached. At that point the cycle is repeated.

Both the Nixie and relay control decoders receive BCD data from the decade counters. These counters are advanced one address or count at a time by pulses from the clock. The decades can be advanced as well to a given count by pulses applied through the first-address switches, when there is a strobe input. All of the decade counters are set to zero when there is a reset pulse.

When the equipment is first turned on, a short-duration reset pulse is applied, together with long-duration strobe and address pulses (Fig. 4–27). The reset pulse sets the decade counters to zero. The strobe pulse enables the counters so that the first-address pulses can be applied. The counters then advance to the first address. When the scan is started (automatically or by a front panel switch), the clock pulses are applied, advancing the counters one step at a time. This count is converted to a BCD code and is applied to the Nixie and relay control decoders, which produce corresponding decade outputs.

When the counters reach the last address, a pulse will appear simultaneously on all three of the last-address switches. These pulses operate the last-address AND gate. The output of this gate is applied to other circuits (not shown) which initiate the reset and start functions over again. Thus, the channels are scanned continuously between the first and last address.

The function of the false code detect (FCD) gates is to prevent two channels from being addressed simultaneously. When any number greater than nine is generated by a decade counter, the FCD gates for that counter drive the false code detect line true. After an 8-μsec delay, the FCD line resets the decade counters to 000.

4–3.1.3. Self-check Information

The following partial self-check procedure is taken from the Hewlett-Packard manual. Note that the self-check procedures involve the three basic elements of troubleshooting: operation of the controls in a given sequence, analysis of symptoms, and measurement of circuit conditions at test points.

CHECKOUT AND TROUBLESHOOTING DIAGRAM

START

Turn on POWER:

  OK: Nixies indicate 000.

            Rear panel fan operates.

  BAD: Nixies remain off.

            Check a-c power connections.

            Check rear panel fuse.

            Check 115/230 VAC slide switch.

            Check d-c voltage at plus side of A26C1 for 190 VDC ± 10 per cent.

  BAD: Nixies show blurred display.

            Check fuse A26F1, Press RESET after replacement.

  BAD: Nixies indicate number other than 000.

            Press RESET

            Turn POWER off and on again.

            Check FIRST-ADDRESS switches for setting of 000.

            Using oscilloscope, monitor test point A22TP1; turn POWER off and on to observe pulse; negative-going, 4.5 V to 0 V, 0.4 msec. If pulse is absent, trouble is probably on card A22; if present, trouble is probably on card A17.

Press RESET

  OK: Nixies display 001.

  BAD: Nixie display remains at 000.

            Monitor A21TP2 for negative-going 10-μsec 4.5-V pulse each time STEP is pressed. If pulse is absent, trouble is probably on card A19 or card A21.

            Monitor A17TP1; press STEP repeatedly; level shift of 3.5 V occurs on eighth and tenth pressing. If level shifts are absent, trouble is probably in the Units decade counter of A17. If shifts are present, trouble is probably on card A12.

Press RESET again.

  OK: Nixies reset to 000.

  BAD: Nixies remain at 001.

            Monitor A22TP1 while pressing RESET button; negative-going 4.5-V pulse occurs (contact closure). If pulse does not occur, trouble is probably on card A18. If pulse occurs but Nixies do not reset, trouble probably is on card A22.

Press START:
   OK: CONTACTS CLOSED, lamp lights.
   BAD: CONTACTS CLOSED, lamp does not light.
        Monitor A22TP1 while alternately pressing START and RESET
        buttons; 5-V shift should occur. If shift is present, lamp or lamp-
        driver circuit is probably bad. If shift is absent, trouble is proba-
        bly on card A21.

4–4. EXAMPLES OF DIGITAL EQUIPMENT TROUBLE-SHOOTING

        Practical digital troubleshooting is a combination of detective
work or logical thinking, and step-by-step measurements. Thus far in this
chapter, we have described what test equipment is available for digital work,
how to use the equipment effectively in testing digital circuits, and how to
use service literature effectively. The following paragraphs describe the final
step: combining all of these practical techniques with logical thinking to solve
some problems in digital troubleshooting.

   Two circuits are discussed, one very simple (the half-adder) and one more
complex (a complete decade counter and readout). All three aspects of
digital troubleshooting are included: self-check, pulse measurement, and
logical thinking.

4–4.1. Half-Adder Troubleshooting

        Figure 4–28 shows the diagram and truth table for a half-adder
circuit. Assume that this circuit is made up of replaceable gates mounted on
a plug-in printed circuit card and is part of a computer. That is, the entire
half-adder circuit can be replaced as a unit by replacing the plug-in card (as
a field-service measure to get the computer operating immediately). Then
the card can be repaired by replacing the defective gates (as a factory or shop
repair procedure).

   Further assume that the computer failed to solve a given mathematical
equation during self-check, the trouble is localized to the half-adder card, the
card is replaced, and the computer then performs its function normally. This
definitely isolates the problem to the half-adder card.

   Now assume that the card can be connected to a power source (to energize
the gates) and that pulses of appropriate amplitude and polarity can be
applied to the two inputs (addend and augend). The sum and carry output
as well as any other points in the circuit can be monitored with an oscilloscope or logic probe.

   Some service shops have special test fixtures that mate with printed circuit

Addend (Digit A)	Augend (Digit B)	Sum	Carry
O	O	O	O
O	1	1	O
1	O	1	O
1	1	O	1

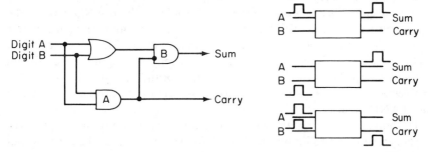

Figure 4–28. Half-adder troubleshooting.

cards. This provides a mount for the card and ready access to the terminals. In other cases, cards are serviced in the equipment by means of an extender. The card is removed from its socket, the extender is installed in the empty socket, and the card is installed on the extender. This maintains normal circuit operation but permits access to the card terminals and components on the card. These arrangements for printed circuit cards are shown in Fig. 4–29.

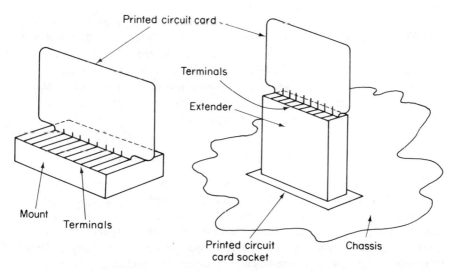

Figure 4–29. Printed circuit card mount and extender.

To test the half-adder of Fig. 4–28, inject a pulse (true) at the addend input (digit A) and check for a true (pulse) condition at the sum output, as well as a false (no pulse) condition at the carry output.

If the response is proper, both the OR gate and the B AND gate are functioning normally. To confirm this, inject a true (pulse) at the augend input (digit B) and check for a true at the sum output, as well as a false condition at the carry output.

If the response is not proper, either the OR gate or the B AND gate are the logical suspects. The A AND gate is probably not at fault, since it requires two inputs (digits A and B) to produce an output. To localize the problem further, inject a pulse at either addend or augend inputs and check for an output from the OR gate. If there is no output or the output is abnormal, the problem is in the OR gate. If the output is normal, the problem is likely in the B AND gate.

To complete the test of the half-adder circuit, inject simultaneous pulses at the addend and augend inputs and check for a true (pulse) at the carry output and a false (no pulse) at the sum output. If the response is proper, all of the gates can be considered as functioning normally. If the response is not correct, the nature of the response can be analyzed to localize the fault.

For example, if the carry output is false (no matter what the condition of the sum output), the A AND gate produces a true output to the carry line when there are two true inputs. Since the test is made by injecting two true inputs, a false condition on the carry line points to a defective A AND gate. A possible exception is where there is a short in the carry line (possibly in the printed wiring). This can be checked by removing all power and checking the resistance of the carry line to ground or common.

As another example, if the carry output is normal (true) but the sum output is also true, the B AND gate is the most likely suspect. The A AND gate produces a true output when both inputs are true. The B AND gate requires one true input from the OR gate and the false input from the A AND gate to produce a true condition at the sum output. If the input to the B AND gate from the A AND gate is true but the sum output shows a true condition, a defective B AND gate is indicated.

4–4.2. Decade Counter and Readout Troubleshooting

Figure 4–30 shows the logic diagram for a three-digit decade counter and readout. Each digit is displayed by means of a separate Nixie tube. Each tube is driven by a separate 4/10-line decoder and storage IC package. The three decoder/storage packages are enabled by a clock pulse at regular intervals or on demand. Each decoder/storage unit receives BCD information from a separate decade counter IC package. The counter packages contain four FFs and produce a BCD output that corresponds to

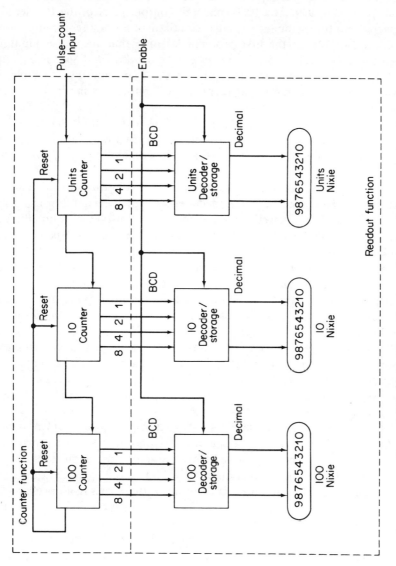

Figure 4-30. Logic diagram for three-digit decade counter and readout.

195

the number of input pulses occurring between reset pulses. The maximum readout possible is 999. At a 1000 count, the output pulse of the 100 decade is applied to all three counter packages simultaneously as a reset pulse.

Problem No. 1. For the first problem, assume that the pulse input is applied through a gate and that the gate is held open for 1 sec. The count shown on the Nixie readout then indicates the frequency. For example, if the count is 377, this shows that there are 377 pulses passing in 1 sec; the frequency is 377 Hz.

Assume that a 700-Hz pulse train is applied to the input. The counter should then go from 000 to 700. However, assume that the count is 000, 001, 002, 003, 004, 005, 000, 001, 002, 003, and so on. That is, the count never reaches 006.

One logical conclusion here is that the problem is in the counter function rather than in the readout function (decoder/storage or Nixie). For example, if the Units readout is defective so that there is no display beyond 005, the 10 and 100 readouts will be unaffected. That is, the count will be 005, 010, 011, 012, and so on.

With the problem localized to the counter function, there are three logical possibilities: the input pulses never reach more than five, (not likely); there is a reset pulse occuring at the same time as the sixth input pulse or any time after the fifth input pulse (possible); and the Units counter simply does not count beyond five, (most likely).

With the detective work out of the way, the next practical step is to make measurements. To confirm or deny the reset pulse possibility, monitor the input pulse line and the reset line of the Units counter on a dual-trace oscilloscope. Adjust the oscilloscope sweep frequency so that about ten input pulses are displayed.

If a reset pulse does occur after the fifth input pulse, the problem is pinpointed. Trace the reset line to the source of the unwanted (improperly timed) reset pulse. If there is no reset pulse before the sixth pulse, monitor the input line and the 4 line of the Units counter. The 4 line should go true at the sixth input pulse (as should the 2 line). Then monitor the 8 line. In all probability, the Units counter is defective: the 4 and 2 lines will show no output, or the output will be abnormal.

If a logic clip is available, check the operation of the Units decade counter on a real-time basis. That is, inject input pulses and check that the 4 and 2 lines go true when the sixth input pulse is injected. If circuit conditions make it possible, disconnect the 4 and 2 lines and recheck operation of the Units counter. It is possible that the 4 or 2 lines are shorted or otherwise defective. Unfortunately, with most present-day logic circuits, the lines or wiring to the IC modules are in the form of printed circuits, making it impractical to disconnect individual lines or leads. The entire IC must be checked by substitution.

Problem No. 2. Assume that the test conditions are the same as for Problem 1. However, the readout is 000, 001, 002, 003, 004, 005, 006, 007, 000, 001, 010, 011, 012, and so on. That is, the 008 and 009 displays are not correct.

The logical conclusion here is that the problem is in the readout function rather than in the counter function. For example, if there is a failure in the counter, the 10 counter will never receive an input from the Units counter. Since there is some readout from the 10 counter, it can be assumed that the Units counter is functioning.

Assuming that the readout is faulty, there are several logical possibilities: the Units counter output lines can be shorted or broken (thus, the Units decoder receives no input or an abnormal input); the Units decoder can be defective; or the Units Nixie tube is defective.

The first practical step depends on the test equipment available. Monitor the 8 and 9 lines to the Nixie tube with a logic probe or oscilloscope. If pulses are present but there is no 8 or 9 display, the Nixie tube is at fault. As an alternative first step, inject a pulse at the 8 and 9 lines of the Nixie and check for a proper display.

If there is no 8 or 9 pulse present at the Nixie input, check for an 8 or 9 input to the Units decoder. An 8 input is produced when there is a pulse at the 8 line (between counter and decoder). A 9 input requires simultaneous pulses on the 8 and 1 lines.

If the 8 and 9 inputs are available to the decoder but there are no 8 and 9 pulses to the Nixie inputs, the Units decoder is at fault. If a logic clip is available, check the operation of the Units decoder on a real-time basis. That is, inject input pulses at the 8 line (from the counter), while simultaneously enabling the decoder, and check that the 8 output line (to the Nixie) goes true.

Problem No. 3. Assume that the test conditions are the same as for Problem 1, except that the input frequency is 300 Hz. However, the readout is 600. That is, the readout is twice the correct value. The logical conclusion here is that the problem is in the counter function rather than in the readout. It is possible that the readout could be at fault, but it is not likely.

A common cause for faults of this type (where FFs are involved) is that one FF is following the input pulses directly. That is, the normal FF function is to go through a complete change of states for two input pulses. Instead, the faulty FF is changing states completely for each input pulse (similar to the operation of an OS). Thus, the decade counter containing such a faulty FF produces two output pulses to the next counter for every ten input pulses, or the counter divides by 5 instead of 10. All decades following the defective stage receive two input pulses, where they should receive one.

The first practical step is to monitor the input and output of each decade in turn. The decade that shows one output for five inputs is at fault.

Problem No. 4. Assume that the test conditions are the same as for Problem 1. However, the readout is 000, 000, 000, 000, and so on. That is, the Nixie tubes glow but remain at 000.

The logical conclusion here is that the Nixies are receiving power and are operative. If not, the Nixies could not produce a 000 indication. The most likely causes for such a symptom are no input pulses arriving at the units counter, a simultaneous reset pulse with the first input pulse (or a short on the reset line), no enable pulse to the decoder/storage packages, or a defective Units counter (not responding to the first input pulse).

Here the practical steps are to monitor (simultaneously) the input line and reset line, input line and enable line, and input line and the output of the Units counter. This should pinpoint the problem. For example, if there are no input pulses (or the wiring is defective, possibly shorted, so that the input pulses never reach the Units counter input), there will be no output. If the input pulses are present but a reset pulse arrives simultaneously with the first input pulse, there can be no readout. If there is no enable pulse, there will be no readout even with the counters operating properly. That is, the counters will produce the correct BCD output which is then stored in the decoders. However, the absence of an enable pulse will prevent the stored information from being displayed on the Nixies. If the input pulses are present and there is no abnormal reset pulse, the output of the Units counter should show one pulse for ten input pulses. If not, the Units counter is defective.

5. TROUBLESHOOTING LABORATORY AND INDUSTRIAL EQUIPMENT

This chapter deals with practical troubleshooting techniques for specialized solid-state circuits, particularly those found in laboratory and industrial equipment. Defects in most solid-state equipment can be located by following the methods described in Chapters 1 and 2. Therefore, these chapters and the procedures applied to all troubleshooting problems should be studied thoroughly.

In addition, there are certain troubleshooting approaches and techniques that can be applied to specialized circuits for rapid localization of problems. This chapter deals with these approaches from a practical standpoint.

It is assumed that the reader is already familiar with the basic principles of these circuits, although the reader may not be familiar with the solid-state version. Therefore, there is no detailed discussion concerning the overall function of the circuit. Instead, to provide a good foundation for troubleshooting, there is a brief discussion of the operational theory for each solid-state circuit given.

A consistant format is used throughout the chapter for each circuit. First, a typical solid-state version of a laboratory or industrial circuit is given along with the basic theory of operation. Next, the failure patterns for that particular type of circuit are discussed. If applicable, a group of typical troubles are described along with the most likely causes of such troubles.

Much of the information in this chapter is based on data gathered by the product training staff of Hewlett-Packard.

199

5–1. BASIC FEEDBACK AMPLIFIERS

Feedback amplifiers are used in a great variety of solid-state laboratory and industrial equipment. Likewise, there are many variations of the basic feedback circuit. Because of this, we shall not deal with a specific feedback circuit but rather with feedback circuits in general, giving typical examples and general approaches.

When troubleshooting any feedback amplifier, there is always a question of approach. Obviously, input and output waveforms can be measured with an oscilloscope, followed by voltage measurements at all transistor elements. This will pinpoint many troubles.

However, such problems as measurement of gain can be a particular concern. For example, if you try opening the loop to make gain measurements, you usually find that there is so much gain that the amplifier saturates and the measurements are meaningless. On the other hand, if you start making waveform measurements on a working, closed-loop system, you often find the input and output signals are normal, while inside the loop many of the waveforms are distorted. For this reason, feedback loops require special attention.

Figure 5–1 is the schematic of a basic feedback amplifier. Note the various waveforms around the circuit. These waveforms are similar to those that will be seen if the amplifier is used with sinewaves. Note that there is an approximate 15 per cent distortion inside the feedback loop (between Q_1 and Q_2) but only a 0.5 per cent distortion at the output. This is only slightly greater

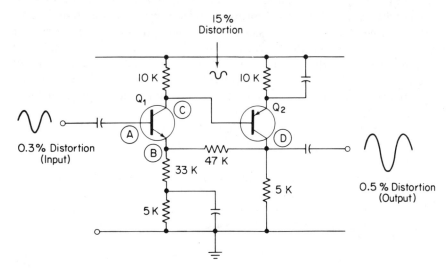

Figure 5–1. Basic feedback amplifier.

distortion than that at the input, which is 0.3 per cent. Open-loop gain for this circuit is approximately 4300, while closed-loop gain is approximately 1000. The gain ratio (open loop to closed loop) of 4 to 1 is typical for feedback amplifiers used in laboratory circuits.

Transistors in feedback amplifiers behave just like transistors in any other circuit. That is, the transistors respond to all of the same rules for gain and input/output impedance. Specifically, each transistor amplifies the signal appearing *between its emitter and base*. It is here that the biggest difference occurs between gain stages in feedback amplifiers and gain stages in non-feedback (open-loop) feedback amplifiers.

Transistor Q_1 in Fig. 5–1 has a varying signal on *both* the emitter and base. In a nonfeedback amplifier the signal usually only varies at one element, either the emitter or the base. Since most feedback systems use negative feedback, the signals at both the base and the emitter are in phase. The resultant gain is much less than when one of these elements is fixed (no feedback, open loop).

This accounts for the great amplifier gain increase when the loop is opened. Either the base or the emitter of the transistor stops moving and the base-emitter control elements see a much larger effective input signal. Assume that a perfect signal is applied to the input (point A of Fig. 5–1). If the amplifier is perfect (introduces no distortion), the signal returning to B will also be undistorted. Since the system uses negative feedback, the signal that travels around the loop a second time will be undistorted as well. If the amplifier is not perfect (assume an extreme case of clipping distortion), the returning signal will show the effect of distortion, as in Fig. 5–2.

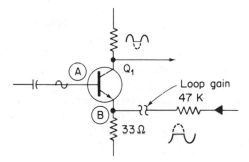

Figure 5–2. Amplifier-induced distortion in signal returning to point B.

To simplify the explanation assume that the clipping is introduced in Q_1 and that Q_2 is perfect. Now the signals applied to the base and emitter of Q_1 are not identical. The resultant applied signal at the control point of Q_1 will be quite distorted. In effect, the distortion will be a mirror image of the distortion introduced by Q_1. Transistor Q_1 then amplifies this distortion and adds in its own counter distortion. The result then, after many trips around the loop, is that there can be distortion inside the loop, but it is counterbalanced by the feedback system. The final output is undistorted or relatively free of amplifier-induced distortion. The higher the amplification and the greater the feedback, the more effective this cancellation becomes and the lower the output distortion becomes.

This last fact marks the basic difference in troubleshooting a feedback amplifier. In any amplifier there are three basic causes of distortion: overdriving, operating the transistor at the wrong bias point, and the inherent nonlinearity of any solid-state device.

Overdriving can be the result of many causes (too much input signal, too much gain in the previous stage, etc.). However, the net result is that the output signal is clipped on one peak due to the transistor being driven into saturation and on the other peak by driving the transistor below cutoff.

Operating at the wrong bias point can also produce clipping but of only one peak. For example, if the input signal is 1 V and the transistor is biased at 1 V, the input will swing from 0.5 V to 1.5 V. Assume that the transistor will saturate at any point above 1.6 V and will cut off at any point below 0.4 V. Further assume that the bias point is shifted (due to component aging, transistor leakage, etc.) to 1.3 V. The input will now swing from 0.8 V to 1.8 V, and the transistor will saturate when one peak goes from 1.6 V to 1.8 V. If the bias point is shifted down to 0.7 V, the input will swing from 0.2 V to 1.2 V, and the opposite peak will be clipped as the transistor goes into cutoff. Even if the transistor is not driven into saturation or cutoff, it is still possible to operate a transistor on a nonlinear portion of its curve due to wrong bias. All transistors have some portion of their input/output curve that is more linear than other portions. That is, the output increases (or decreases) directly in proportion to input. An increase of 10 per cent at the input produces an increase of 10 per cent at the output. Ideally, transistors are operated at the center of this linear curve. If the bias point is changed, the transistor can operate on a portion of the curve that is less linear than the desired point.

The *inherent nonlinearity* of any solid-state device (diode, transistor) can produce distortion even if a stage is not overdriven and is properly biased. That is, the output will never increase (or decrease) directly in proportion to input. For example, an increase of 10 per cent at the input can produce an increase of 13 per cent (or 7 per cent) at the output. This is one of the main reasons for feedback in amplifiers where low distortion is required.

In summary, a negative feedback loop operates to minimize distortion in addition to stabilizing gain. The feedback takeoff point will have the minimum distortion of any point within the loop. From a practical troubleshooting standpoint, if the *final output* distortion and the *overall gain* are within limits, all of the stages within the loop can be considered as operating properly. Even if there is some distortion or abnormal gain in one or more of the stages, the overall feedback system has compensated for the problem. Of course, if the overall gain and/or distortion are not within limits, the individual stages must be checked.

5-1.1. Failure Patterns for Feedback Amplifiers

Most feedback amplifier problems can be pinpointed by waveform measurement and the basic solid-state troubleshooting techniques described in Chapter 2 (voltage/resistance measurements, shorting transistor elements to see effect on gain and leakage, etc.). The following points should be given special attention when troubleshooting any feedback amplifier circuit.

Opening the loop. Some troubleshooting literature recommends that the loop be opened and the circuits checked under no-feedback conditions. In some cases, this can cause circuit damage. Even if there is no damage, the technique is rarely effective. Open-loop gain is usually so high that some stage will block or distort badly. If the technique is used as it must be for some circuits (typically an operational amplifier), keep in mind that distortion will be increased. That is, a normally closed-loop amplifier can show considerable distortion when operated as an open loop, even though the amplifier is good.

Measuring stage gain. Care should be taken when measuring the gain of amplifier stages in a feedback amplifier. For example, in Fig. 5-1 if you measured the signal at the collector of Q_1 and divided this by the signal from the Q_1 transistor base to ground, you will have the wrong answer. The reason is that gain or amplification is defined as the output voltage divided by the input voltage: voltage/gain $= V_o/V_{in}$.

By definition, the input voltage is the voltage applied to the transistor control elements (the voltage between the emitter and base). Since a feedback signal is applied to the emitter of Q_1, the base-to-ground voltage will not be the same as the input voltage. To get the correct value, connect the low side of the measuring device (a-c voltmeter or oscilloscope) to the emitter and the other lead to the base (See Fig. 5-3). In effect, measure the signal across the base-emitter junction. This will include the effect of the feedback signal.

As a general safety precaution, never connect the ground lead of a voltmeter or oscilloscope to the base of a transistor, unless that lead connects back to an *isolated inner chassis.* The reason is because large a-c ground loop currents can flow through the base-emitter junction (and then to ground) and easily blow out the transistor.

Low-gain problems. As discussed, low gain in a feedback amplifier can also result in distortion. That is, if gain is normal in a feedback amplifier, some distortion can be overcome. With low gain, the feedback may not be able to bring the distortion within limits. Of course, low gain by itself is sufficient cause to troubleshoot any feedback amplifier.

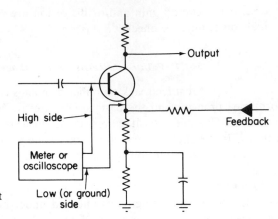

Figure 5–3. Measuring input signal voltage or waveform.

Take the classic failure pattern of a solid-state feedback amplifier that was working properly, but now the output is low by about 10 per cent. This indicates a general deterioration of performance rather than a major breakdown.

Keep in mind that most feedback amplifiers have a very high open-loop gain that is set to some specific value by the ratio of resistors (feedback resistor to load resistor). If the closed-loop gain is low, it usually means that the open-loop gain has fallen far enough so that the resistors no longer determine the gain. For example, if the a-c of beta of Q_2 in Fig. 5–1 is lowered, the open-loop gain will be lowered. Also, the lower beta will lower the input impedance of Q_2 which, in turn, will reduce the effective value of the load resistor for Q_1. This also has the effect of lowering overall gain.

In troubleshooting such a situation, if waveforms indicate low gain and element voltages are normal, try replacing the transistors. Of course, never overlook the possibility of open or badly leaking emitter-bypass capacitors. If the capacitors are bad, there will be considerable negative feedback and little or no a-c gain.

Distortion problems. As discussed, distortion can be caused by improper bias, overdriving (too much gain), or underdriving (too little gain, preventing the feedback signal from countering the distortion). One problem often overlooked in a feedback amplifier with a distortion failure pattern is overdriving due to transistor leakage.

Generally, it is assumed that collector-base leakage will reduce gain, since the leakage is in opposition to the signal current flow. While this is true in the case of a single stage, it may not be true where more than one feedback stage is involved.

Whenever there is collector-base leakage, the base will assume a voltage nearer to that of the collector (nearer than is the case without leakage). This increases transistor forward bias and increases transistor current flow. An increase in the transistor current causes a lower h_{ib} (a-c input resistance, grounded base configuration) which causes the stage gain to go up. At the same time, a reduction in h_{ib} causes a reduction in common emitter input resistance which may or may not cause a gain reduction (depending on where the transistor is located in the amplifier).

If the feedback amplifier is direct coupled, the effects of leakage are increased. This is because the operating point (base bias) of the following stage is changed, possibly resulting in distortion. For example, the collector of Q_1 is connected directly to the base of Q_2. If Q_1 starts to leak (or the collector-base leakage increases with age), the base of Q_2 (as well as the collector of Q_1) will shift in no-signal voltage.

5–2. BASIC OPERATIONAL AMPLIFIERS

Operational amplifiers are the basic building blocks of analog computers because they can perform the mathematical operations of amplification, addition, subtraction, integration, and differentiation. In combination with nonlinear elements such as solid-state diodes, operational amplifiers can be used as limiters, level detectors, as well as nonlinear function generators.

For these reasons, the operational amplifier is found in much laboratory and industrial equipment. Often, the operational amplifier appears in the form of an IC package, since such packages are ideally suited to the requirements for an operational amplifier. The basic operational amplifier has a differential input followed by several high-gain stages, all of which can be combined into an IC package. Feedback, which is required for all operational amplifier configurations, is external to the package. This makes it possible to use the same IC package for many operational amplifier functions simply by selecting different external feedback components. Of course, solid-state operational amplifiers are also found in printed circuit card form.

From a practical troubleshooting standpoint, it is often necessary to service operational amplifiers by working with the external feedback components. In the case of IC packages, the external components are the only ones that can be tested or replaced. Even with printed circuit card operational amplifiers, troubleshooting starts by isolating the problem to the external components or the amplifier components. That is, the amplifier is tested as a separate function first. If the amplifier performs properly, the trouble is isolated to the external components, and vice versa.

To service equipment containing an operational amplifier properly, it is necessary to understand the relationship of external components to the basic amplifier. To aid in this understanding, the following sections describe some typical operational amplifier configurations.

5–2.1. Typical Operational Amplifier Configurations

An operational amplifier is a very high-gain, (usually) direct-coupled amplifier having out-of-phase input/output characteristics (180°

phase shift). It is possible to convert a differential amplifier, such as is discussed in Sec. 5–4, to an operational amplifier by the addition of a feedback network. Most IC operational amplifiers have a differential input. Since the open-loop gain (without feedback) is very high, closed-loop (with feedback) characteristics can be controlled by feedback components within gain limits of the amplifier.

Normally, resistors and capacitors are used as input and feedback components. By selecting proper feedback networks, many functions can be performed. From a troubleshooting standpoint, failure of equipment (using operational amplifiers) to perform its overall function (integration, amplification by a constant, etc.) is usually the result of a problem in the external components. On the other hand, if the overall function is performed properly but there are problems such as excessive noise, drift, and so on, the fault lies in the amplifier.

Figure 5–4 is the generalized or theoretical feedback circuit and gain equation for an operational amplifier. Note that both a positive and negative input are shown. Signals applied to the negative input are inverted at the output, while signals at the positive input arrive at the output with the same polarity. The negative (inverting) input is normally used in an operational amplifier because the inverted output permits the use of negative feedback through the Z_f feedback components, generally referred to as the external components. When not in use, the plus input is usually grounded. Therefore,

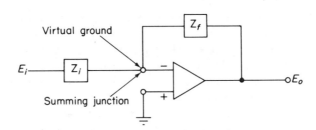

$$G = \frac{E_o}{E_i} = -\frac{Z_f}{Z_i}\left[\frac{1}{1 - 1/A\,(1 + Z_f/Z_i)}\right]$$

G = closed−loop gain
E_o = output voltage
E_i = input voltage
Z_f = feedback impedance
Z_i = input impedance
A = open−loop gain

Figure 5–4. Generalized or theoretical feedback circuit and gain equation for operational amplifier.

the output will be related to the difference between the negative input and ground rather than a difference between two inputs.

As shown, the theoretical gain for an operational amplifier depends on Z_f (feedback impedance), Z_i (input impedance), and A (open-loop amplification of the amplifier).

When the positive input is grounded, a concept of *virtual ground* is applied to the negative input. Actually, the d-c level at the negative input of an operational amplifier is very close to ground. When an input signal is applied, the signal tends to move the base away from ground. However, the negative feedback from the output of the amplifier resists this tendency. The amount that the negative input voltage varies with a signal depends on the open-loop gain of the amplifier; the higher the gain, the less the negative input voltage varies. This is the same for any feedback amplifier, as described in Sec. 5–1. (An increase in feedback decreases distortion.)

With the high open-loop gain normally found in an operational amplifier, the negative input voltage varies only slightly under closed-loop conditions. It is convenient to assume that for all practical purposes the negative input voltage does not change with signal. Thus, is appears as though the negative input is grounded. The term *virtual ground* is used to indicate that, although the input of the amplifier appears to be grounded, actually it is not. (Many equations for the functions performed by an operational amplifier can be derived most easily by the use of the virtual ground concept.)

It should also be noted that since a virtual ground exists at the negative input, the input impedance of the amplifier is essentially determined by the value of the Z_i component. For example, when an operational amplifier is used in laboratory test equipment, the input component is a 50-Ω resistor. Thus, the source sees 50 Ω, no matter what signal level is applied to the input.

5–2.1.1. Amplification by a constant (fixed gain)

When you want to amplify a signal by a constant (that is, to provide a fixed gain of 100, 1000, and so on), the circuit in Fig. 5–5 is used.

Any desired gain can be obtained with such a circuit, provided the gain is within the open-loop gain and frequency limits of the amplifier. The closed-loop gain is determined by the feedback resistance values. In effect, the input voltage will be multiplied by the ratio of the feedback resistance R_f to input resistance R_i. For example, if R_i is 1000 and R_f is 10,000, then a gain of 10 will be obtained. If a 0.5-V input is applied, the output will be 5 V. If the output is capable of a 10-V swing, an input signal up to 1 V can be applied without overdriving the amplifier. (As is discussed in later sections, the fact that closed-loop gain can be controlled by the ratio of R_i and R_f is used as a troubleshooting device for any operational amplifier.)

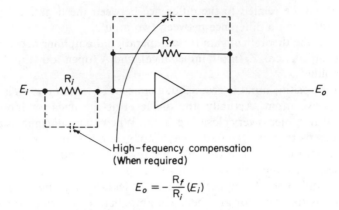

High-fequency compensation
(When required)

$$E_o = - \frac{R_f}{R_i}(E_i)$$

Figure 5–5. Amplification by a constant.

The output of the circuit in Fig. 5–5 is inverted. Thus, the circuit can serve as a sign changer with or without amplification. For example, if the circuit is used to amplify a 1-mV positive pulse and the resistance values are selected to provide a gain of 10, the output will be negative ($-$) 10-mV pulses.

The circuit can be made to provide variable gain instead of gain by a fixed constant, if the feedback resistor R_f is a potentiometer. Likewise, feedback resistance R_f can be replaced by thermistors, photoresistors, or other variable resistance elements. The gain is then a function of the temperature, light level, or other variable.

5–2.1.2. Integration of Signals

When it is desired to integrate signals, the circuit of Fig. 5–6 is used. (Since operational amplifier integrators are in such common use, Sec. 5–3 is devoted to their special troubleshooting problems.)

Note that the input circuit of Fig. 5–6 is formed by series resistance R_i, while feedback is accomplished with a fixed capacitor C_f. With such an arrangement, the output voltage is directly proportional to the integral of the input voltage and inversely proportional to the time constant of the feedback network (R_iC_f). The time constant is *approximately equal* to the period of the signal to be integrated.

The circuit of Fig. 5–6 can be used as a precision 90° *phase shifter*. Sinewaves applied to the input of the integrator are shifted in phase by exactly 90°. Sinewaves of any frequency can be applied, provided they are within the frequency limits of the amplifier.

One problem often encountered in practical applications is that the circuit will integrate *all signals* present at the input. That is, the output will represent the integral of all inputs (a-c, d-c, drift noise, etc.). This can cause a problem

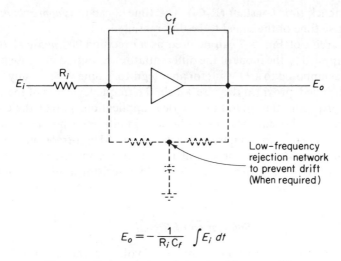

$$E_o = - \frac{1}{R_i \, C_f} \int E_i \, dt$$

Figure 5-6. Integration of signals with an amplifier

when you want to integrate only the a-c signals (the usual case). For example, if the output is displayed on an oscilloscope and d-c is present with the a-c signals at the input, the output will gradually increase, resulting in drift of the oscilloscope display. Some circuits use a low-frequency rejection network similar to that of Fig. 5-6 to prevent such output drift.

5-2.1.3. Differentiation of Signals

When it is desired to differentiate signals, the circuit of Fig. 5-7 is used. Note that the input circuit is formed by a series capacitor C_j, while the feedback is accomplished with a fixed resistor R_f. With such an arrangement, the output voltage is directly proportional to the time rate of change (or frequency) of the input voltage and is inversely proportional to

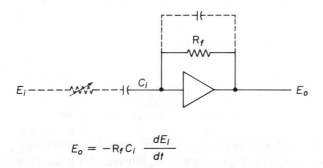

$$E_o = -R_f C_i \, \frac{dE_I}{dt}$$

Figure 5-7. Differentiation of signals with an amplifier.

the feedback time constant ($R_f C_i$). The time constant is *approximately equal* to the rise time of the signal to be differentiated.

The circuit of Fig. 5–7 can be used as a precision 90° *phase shifter*. Sinewaves applied to the input of the differentiator are shifted in phase by exactly −90° (as opposed to a 90° shift for an integrator). Sinewaves of any frequency can be applied, provided they are within the frequency limits of the amplifier.

One problem often found in practical applications is that the circuit will accentuate high-frequency noise. Some circuits incorporate a noise suppression network to minimize this condition. High-frequency noise suppression in a differentiation circuit is usually accomplished by means of a capacitor across the feedback resistor or by a resistor in series with the input capacitor, as shown in Fig. 5–7.

5–2.1.4. Summation of Signals

When it is desired to sum a number of voltages, the circuit of Fig. 5–8 is used. This is typical of the circuits used in solid-state analog computers, where the operational amplifier is an IC. Note that the input circuit is formed by a number of parallel resistors, one for each voltage to be summed, while the feedback is accomplished by a single resistor.

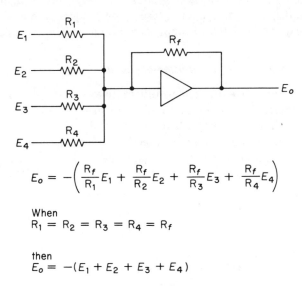

$$E_o = -\left(\frac{R_f}{R_1} E_1 + \frac{R_f}{R_2} E_2 + \frac{R_f}{R_3} E_3 + \frac{R_f}{R_4} E_4 \right)$$

When
$$R_1 = R_2 = R_3 = R_4 = R_f$$

then
$$E_o = -(E_1 + E_2 + E_3 + E_4)$$

Figure 5–8. Summation of signals with an amplifier.

When the values of all resistors are the same, the output of the amplifier is the sum of all input voltages, with the sign inverted. For example, assume that all of the resistors are 10 Ω (including R_f) and that the voltage across R_1 is

10 V, across R_2 is -7 V, across R_3 is 5 V, and across R_4 is -1 V. Then the output is -7 V, since $10 - 7 + 5 - 1 = 7$; 7 inverted $= -7$

5–2.1.5. Unity Gain with High-Input Impedance

When it is desired to provide a high-impedance input with unity gain, the circuit of Fig. 5–9 is used. Since the input impedance of most solid-state devices (except FETs) is low, the configuration of Fig. 5–9 is used where high impedance is a must.

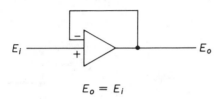

$$E_o = E_i$$

Figure 5–9. Unity gain with high-input impedance.

Note that there are no input or feedback resistors. The feedback is applied directly to the negative input to provide a gain of 1 V at the amplifier output. Since the signal is applied to the positive input, the output is not inverted.

Because the signal is applied directly to the positive input and there are no resistance elements between it and ground, the input impedance of the circuit is determined primarily by the input current. Since this is quite small in most cases, the input impedance is very high. Of course, this is the d-c impedance. For a-c signals, the shunt capacitance brings the impedance down to a much lower level.

5–2.1.6. Amplification with High-Input Impedance

When it is desired to provide a high-input impedance with some gain, the circuit of Fig. 5–10 is used. Note that this circuit is similar to the circuit of Fig. 5–9. The gain is positive (noninverting) and is determined by the equation shown in Fig. 5–10.

In this circuit the signal is applied directly to the positive input with

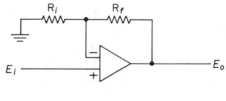

Figure 5–10. Amplification with high-input impedance.

$$E_o = \frac{R_i + R_f}{R_i} E_i$$

feedback from the output to the negative input. The amount of feedback (and the amount of gain) is controlled by the values of R_i and R_f. Note that it is not possible to get a gain of less than 1 with this circuit.

5–2.1.7. Subtraction and/or Difference Amplification

Using an amplifier connected in the circuit of Fig. 5–11, one signal voltage can be subtracted from another by simultaneous application of signals to both inputs of the amplifier. This signal applied to the negative input is subtracted from the signal applied to the positive input.

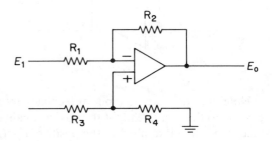

Subtract:

$$E_o = -\frac{R_2}{R_1}E_1 + \left(\frac{R_4}{R_3 + R_4}\right)\left(\frac{R_1 + R_2}{R_1}\right)E_2$$

Difference:
When
$$R_1 = R_2 = R_3 = R_4$$

Figure 5–11. Subtraction and/or difference amplification.

Then
$$E_o = E_2 - E_1$$

5–2.1.8. Voltage-to-Current Converter (Transadmittance Amplifier)

Using an amplifier connected in the circuit of Fig. 5–12, it is possible to supply a current to a load, with the current proportional to the amplifier input voltage. The current supplied to the load is relatively independent of the load characteristics. This circuit is essentially a *current feedback amplifier* and is often identified as such in service literature.

A current-sampling resistor R_s is used to provide the feedback to the positive input. When $R_i = R_f = R_3 = R_4$, the feedback maintains the voltage across current sampling resistor R_s at a value of $-R_f/R \times E_i$, regardless of the load.

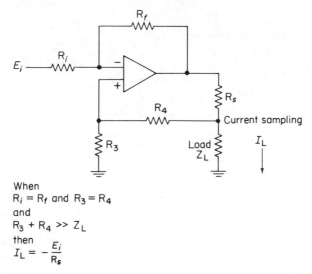

When
$R_i = R_f$ and $R_3 = R_4$
and
$R_3 + R_4 \gg Z_L$
then
$$I_L = -\frac{E_i}{R_s}$$

Figure 5-12. Voltage-to-current converter (transadmittance amplifier).

If a constant input voltage is applied, the voltage across R_s also remains constant regardless of the load. With the voltage across R_s constant, the current through R_s is constant. With R_3 and R_4 normally much higher than the load impedance, the current through the load must remain nearly constant regardless of the impedance.

Normally the values of R_i, R_f, R_3, and R_4 should be the same. The current sampling resistor R_s is then selected for the desired load currents. When R_s is expressed in kilohms, the current through the load $I_L = E_i/R_s$, or milliamperes per volt of input signal.

5-2.1.9. Voltage-to-Voltage Amplifier

Using an amplifier connected in the circuit of Fig. 5-13, it is possible to supply a voltage across a load that will remain constant, regardless of changes in the load. This circuit is similar to that of Fig. 5-12 (voltage-to-current converter) except that the load and current-sensing resistor R_s are transposed. With $R_i = R_f = R_3 = R_4$, the feedback to the negative input maintains the voltage across the load equal to input voltage regardless of load (within the current limitations of the amplifier).

5-2.1.10. Bandpass Amplifier

An operational amplifier can be used as a bandpass amplifier. The basic circuit is shown in Fig. 5-14.

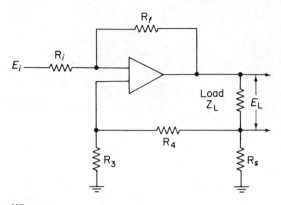

When
$R_i = R_f$ and $R_3 = R_4$
then
$E_L = -E_i$

Figure 5–13. Voltage-to-voltage amplifier (voltage-gain amplifier).

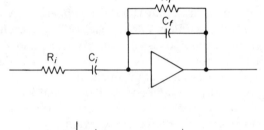

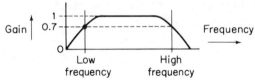

If both R_i and R_f = 159 K then

Figure 5–14. Basic bandpass amplifier.

Low frequency	C_i	High frequency	C_f
1 Hz	1 μF	100 kHz	10 pf
10 Hz	0.1 μF	10 kHz	0.0001 μF
100 Hz	0.01 μF	1 kHz	0.001 μF
1 kHz	0.001 μF	100 Hz	0.01 μF
10 kHz	0.0001 μF	10 Hz	0.1 μF
100 kHz	10 pF	1 Hz	1 μF

In this circuit, the input series R_i and C_i attenuate low frequencies, and the feedback parallels R_f and C_f attenuate high frequencies. If both the input and feedback resistance values are the same, there will be unity gain across the flat portion of the bandpass curve as shown in Fig. 5–14.

The values for C_i and C_f will provide an *approximate* 3 dB/octave drop when the values of both resistors are 159 K. For example, if 0.1 μF is used for C_i and 0.0001 μF is used for C_f, then the output will be 3 dB down from unity gain at 10 Hz and at 10 kHz.

5–2.1.11. *Frequency-to-Voltage Converter*

An operational amplifier can be used to provide a d-c output that is proportional to input frequency. This is the basis for many test equipment circuits. The basic circuit is shown in Fig. 5–15. Note that this is similar to the differentiator circuit of Fig. 5–7. However, the output of Fig. 5–15 is rectified and used to charge a capacitor. Since the output of a differentiator is proportional to frequency, the capacitor charge and the d-c voltage thus obtained are proportional to frequency.

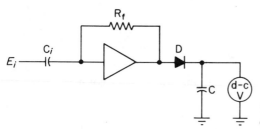

Figure 5–15. Frequency-to-voltage conversion with an amplifier.

5–2.1.12. *Peak-Reading Amplifier*

An operational amplifier can be used to measure the peak voltage of various waveforms. The basic circuit is shown in Fig. 5–16. In

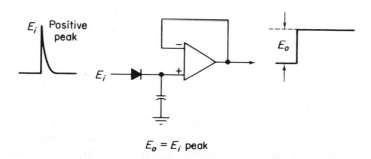

$E_o = E_i$ peak

Figure 5–16. Peak-reading amplifier.

this circuit advantage is taken of the high-input impedance feature of the positive input. When a positive pulse is applied, the diode conducts, charging the capacitor to the peak voltage. Because of the high-input impedance, the capacitor charge is retained for a relatively long time. Under these conditions, unity gain is obtained from the amplifier. Therefore, the output is equal to the peak voltage of the input pulse.

For the circuit to operate properly, the time constant of the source impedance and the capacitor to be charged must be short enough so that the capacitor can charge to the peak voltage in the time that the pulse remains at the peak. Thus, the capacitor value should be as small as possible. The capacitor cannot be too small, however, or it will discharge too rapidly.

A capacitor with a very low leakage should be selected to prevent rapid loss of the charge. Also, the diode reverse current should be very low so as to prevent the capacitor charge from being lost too rapidly. The forward drop across some silicon diodes is great enough to prevent the capacitor from charging to the peak voltage. This is especially true where the input voltage is 1 V or less.

From a practical troubleshooting standpoint, the capacitor and diode are prime suspects when the symptoms are inaccurate readings, or when you cannot obtain a satisfactory calibration.

5-2.2. D-C Stability Problems

Since most operational amplifiers are direct coupled, they are subject to various forms of drift. For example, a d-c amplifier cannot tell the difference between a change in the supply voltage or a change in the signal voltage. Because of the instability problems, most operational amplifiers incorporate a zero-correction circuit, and many amplifiers have a chopper stabilization circuit. Generally, these circuits are external to the operational amplifier.

In theory, the output voltage of an operational amplifier depends on the ratio of input and feedback resistors as well as the input voltage. If the input voltage is zero, the output voltage will be zero. In practice, however, it will be found that there is a small, unwanted drift voltage which appears at the amplifier output terminals, even when the input terminals are short circuits (no signal). With solid-state amplifiers, the offset voltage may amount to tens of millivolts and often comes from transistor leakage currents or static charges.

The undesired voltage (often referred to as zero offset) has both a truly random component (very-low-frequency noise) and a component which is temperature sensitive. The temperature-dependent component can be taken care of easily by injecting an extremely small offset voltage or current into the input circuit which cancels the internally generated thermal offset (or an offset due to poor design). The zero-correction circuit for an operational amplifier is shown in Fig. 5–17.

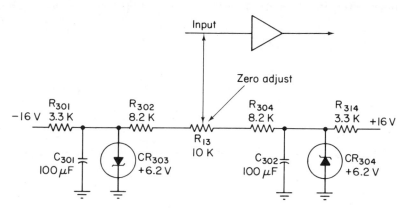

Figure 5–17. Zero-correction circuit for an operational amplifier (Hewlett-Packard).

5-2.2.1. Chopper Stabilization

While the circuit of Fig. 5–17 will compensate adequately for thermal drift and for long-term component changes (aging, etc.) the circuit cannot remove the low-frequency noise which is the random component of the output zero offset. Chopper stabilization is often used to remove this type of drift.

Both electromechanical and optical choppers are used with operational amplifiers. However, the trend is toward optical choppers for solid-state equipment. The basic chopper and amplifier circuit are shown in Fig. 5–18. The d-c signal path is through V_1, C_1, A_1, C_2, V_3, and A_2, while the a-c signal path is via C_3 and A_2. The input signal at C_1 is a squarewave derived from the d-c component of the input signal chopped by the alternate illumination of photoresistors V_1 and V_2. The input signal to the low-pass filter at the output of A_1 contains the amplified d-c component of the input signal recovered by the alternate illumination of photoconductors V_3 and V_4. Typically, the illumination for all of the photoconductors is derived from a bistable neon tube relaxation oscillator.

Note that when V_1 is illuminated, V_2 is dark, and vice versa. When V_1 is illuminated and V_2 is dark, the signal passes and charges C_1. On the alternate cycle, V_1 is dark and offers a high resistance to the signal. Likewise, V_1 is illuminated, allowing C_1 to discharge to ground.

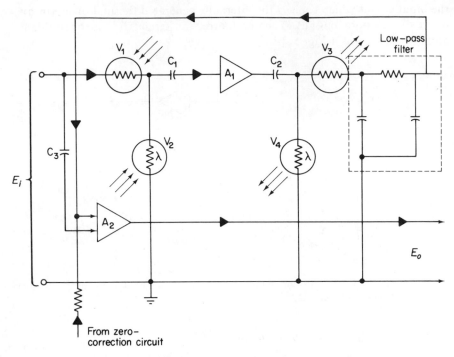

Figure 5–18. Chopper stabilization for operational amplifier.

5–2.3. Overload Protection

Operational amplifiers generally contain considerable overload protection. This prevents damage to the transistors which occurs on overloads due to saturation. The protection also prevents the various a-c coupling capacitors (if used) from charging up. Such capacitors are usually large enough in value to ensure good low-frequency response. If these capacitors become fully charged, the transient recovery time of the amplifier is long. With proper overload protection, transient recovery time can be kept to about 20 μsec for a typical solid-state operational amplifier.

Figure 5–19 is the overload protection circuit for a typical operational amplifier. Input diodes CR_{210} and CR_{211} prevent the input voltage of the summing junction from ever exceeding ± 0.6 V, thus protecting the input transistors from being destroyed. Zener diodes CR_{301} and CR_{302} have nominal breakdown voltages of 10 V, so that whenever the input signal exceeds about ± 11.8 V, either CR_{301} or CR_{302} will exhibit Zener breakdown at 10 V. Feedback current from the output terminal cannot flow back to the summing junction unless one of CR_{314} or CR_{315} and one of CR_{201} or CR_{202}

are also turned on. Thus, when the output voltage reaches 11.8 V (10 + 0.6 + 0.6 + 0.6), large amounts of current from the output terminal are available to oppose the incoming overload current, and thus the entire amplifier is kept from going into saturation. CR_{301} and CR_{302} cannot be used alone since the reverse leakage of these Zener diodes will feed input currents on the order of 1 μA directly into the summing junction. The remaining four diodes (CR_{201}, CR_{202}, CR_{314}, and CR_{315}) and resistors R_{207} and R_{355} shunt the normal Zener diode reverse leakage current to ground.

5–2.4. Failure Patterns for Operational Amplifiers

Major disasters are relatively rare in well-protected operational amplifiers, since input overloads never drive the circuit into saturation. Likewise, when such major failures occur, they are relatively easy to trouble-shoot. That is, the problems are easy to spot by normal signal tracing with waveforms or voltage measurements at the transistor elements. For example, a major failure will usually show up as a normal input but no output, at a particular amplifier stage.

However, operational amplifiers are often plagued with such problems as hum, drift, and noise. The following paragraphs describe the most likely causes for such problems, with practical approaches for locating the faults.

5–2.4.1. Hum and Ripple Problems

In solid-state operational amplifiers, any hum or ripple almost always comes from the d-c power supplies feeding the amplifier. This is unlike vacuum-tube operational amplifiers, where hum or ripple can come from heater-to-cathode leakage. A possible exception is when hum is picked up due to poor shielding or badly grounded leads.

The first step in localizing a hum or ripple problem is to short the input terminals and monitor the output with an oscilloscope. If the hum or ripple is removed when the input terminals are shorted, the hum is probably being picked up by the leads or at the terminal. Look for loose shields, loose ground terminals, and cold solder joints where lead shielding is attached to chassis or feed-through terminals.

If the hum or ripple is not removed when the input terminals are shorted, the hum is probably coming from the power supply. Monitor the power supply voltages at the point where they enter the amplifier. If the power supply is showing an abnormal amount of ripple, the problem is in the power supply. (Refer to Sec. 5–5.) However, since the amplifier has considerable gain, the ripple monitored at the amplifier output may be much greater than at the power supply.

5–2.4.2. Drift and Noise Problems

Drift and noise problems in operational amplifiers are perhaps the most common complaint. There are several places to look in trying to track the causes of noise and drift.

Unstable power supplies. Operational amplifiers are *extremely sensitive* to power supply stability. For example, with solid-state operational amplifiers the typical dual power supply voltages will be ± 12 V or ± 15 V. For satisfactory amplifier operation, the drift should be less than 1 mV/min (or less in some special operational amplifiers used with data processing equipment). Because of the low voltages involved, power supply stability measurements are best made with a five or six-place digital voltmeter. Such a meter can be connected to the monitoring point and checked at least once every minute or over at least a 5-min interval. If the drift is less than 1 mV/min over this time interval, the power supply is probably satisfactory for typical operational amplifier use.

Noisy zero-correction circuits. Another possible source of output noise in operational amplifiers is the zero-correction circuit, which takes voltages from both the positive and negative power supplies and provides a small d-c current to oppose the internally generated offset. (See Fig. 5–17.) If Zener diodes are used to regulate some portion of the zero-correction supply voltage (as is done by CR_{303} and CR_{304} in Fig. 5–17), the Zeners should be checked carefully for drift and noise. Keep in mind that any noise (or other signal) at the zero-correction circuit is injected into the amplifier at the point of highest gain (usually at the first-stage input).

Contaminated printed circuit boards. Another frequent source of output drift is contamination of the printed circuit boards on which operational amplifiers are mounted. This even applies to IC packages. Although the IC package containing the complete operational amplifier is sealed, the summing junction is not sealed. (The summing junctions are usually placed on Teflon standoffs near the IC package on the printed circuit board.)

When you remember that the input current at the summing junction is typically 10^{-11} A, it is easy to see why any contamination from fingerprints (providing leakage paths into the junction) can cause annoying output instability. Operational amplifier circuit boards should be handled as little as possible, and then only while wearing cotton gloves. Great care should be taken never to touch the summing junction terminals with bare hands. Boards suspected of being contaminated should be washed carefully with a clean degreasing solvent and dried with warm dry air. (Never blow them dry with an air hose, as air lines invariably contain oil and water.)

Leakage in the overload protection circuitry. Another place to look for causes of output noise and drift is the overload protection circuitry. If an amplifier has been subjected to repeated serious overloads, there is a possibility of finding high and unstable reverse leakage currents in the overload protection circuits.

In the circuit of Fig. 5–19, the reverse currents can be monitored by measuring the voltages across R_{207} and R_{355}. Substantial voltages across these resistors indicate considerable reverse leakage. If such voltages are measured in an operational amplifier with a noise symptom, try replacing the diodes, one at a time.

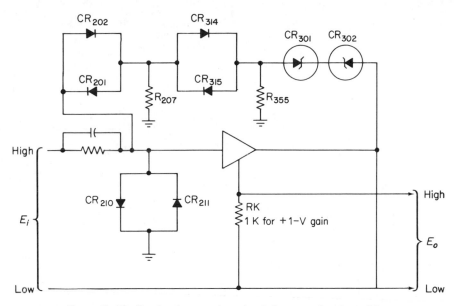

Figure 5–19. Overload protection circuit for operational amplifier.

Unstable choppers. Perhaps the most common source of output instability in chopper-stabilized operational amplifiers is instability of the chopper itself. In the case of electromechanical (vibrator-type) choppers, the problem is essentially one of variations in the dwell time (the time during which the contacts are closed on each half of the cycle). With photoelectric choppers, instabilities in the firing voltage of the neon lamps will produce the same symptoms of erratic output voltages. However, since choppers modulate low-level signals, it is virtually impossible to check choppers except by substitution. Any variation in the musical note of an electromechanical chopper is a sure sign it should be replaced. However, since photoelectric choppers are sealed units, they must be checked by substitution.

5–2.4.3. General Troubleshooting Hints for Operational Amplifiers

Due to their extremely high open-loop gain, troubleshooting operational amplifiers can be quite difficult. The basic test connections are shown in Fig. 5–20. The input is shorted, the drift output (if any) is monitored on a digital voltmeter, and the hum and noise (if any) are monitored on an

Connections for test of hum, drift, and noise

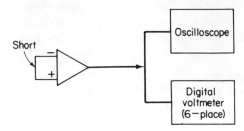

Connections for test of gain, and distortion

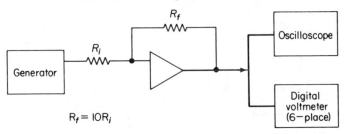

$R_f = 10R_i$

Figure 5–20. Basic troubleshooting connections for operational amplifier.

oscilloscope. If the operational amplifier is used in a circuit where feedback is not by means of a resistor (such as an integrator where the feedback is through a capacitor), a resistor should be inserted in the feedback path. The feedback resistor should be ten times the value of the input resistor. Thus, the amplifier will operate with a gain of 10.

The same test connections can be used to check the range of zero-adjust circuitry and the clipping level of the diode protection circuitry.

To check the zero-correction range, vary the zero control from one end of its range to the other, while observing the amplifier output on the digital voltmeter. (The digital voltmeter should have a sensitivity of at least 1 mV.) If the zero control is provided with steps (instead of or in addition to a variable control), check to see that each step produces the same size step in output voltage. Also check the output voltage stability at each step. If the output voltage appears to be unstable at any particular step, look for poor contacts on the switch, poor solder connections, or a defective resistor connected to the switch contact.

Sometimes it will be found that the range of the zero-adjust control is not sufficient to cause limiting of the amplifier (by means of the protective overload circuitry). If this is the case, apply a small d-c voltage of known stability (preferably from a battery) to the input terminals of the amplifier. Check to see that limiting occurs at the amplifier output at the correct values for *both polarities* of the input voltage.

Unbalance between the positive and negative overload breakdown voltages will suggest either an open or shorted diode in the overload protection loop or saturation taking place internally within the amplifier. In either case, leave the output driven lightly into saturation and measure the d-c voltages appearing across all the elements in the overload protection circuit, plus the operating points of the various stages within the amplifier proper (unless the amplifier is an IC where internal stages cannot be measured).

5–3. BASIC INTEGRATOR (WITH OPERATIONAL AMPLIFIER)

As discussed in Sec. 5–2.1.2, an operational amplifier is often used as an integrator. Figure 5–21 shows a basic circuit using an operational amplifier in a Miller integrator configuration. This is typical of the ICs used in much laboratory equipment.

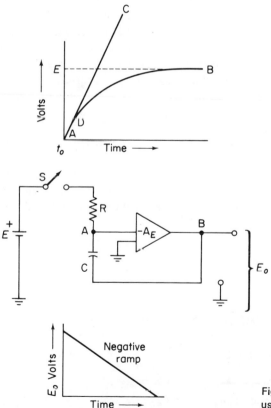

Figure 5–21. Operational amplifier used as Miller integrator.

Notice that the feedback is an amplified and inverted signal from point B (output) to the bottom plate of capacitor C. The reason an RC charging circuit produces an exponential curve is because the charging current through the capacitor is not constant. Instead, the current decreases until no more current will flow.

Current through the capacitor is affected by the *difference in change* on the capacitor plates. The less the difference, the less the current. Current is also affected by the voltage difference across the plates. The more the difference, the more the current. As the capacitor is allowed to charge, current is reduced because the difference in charge across the plate becomes less and less. However, if the voltage difference is increased at the same time. the *net current* flow through the capacitor can be made nearly constant.

This is exactly what the Miller IC of Fig. 5–21 does. When switch S is closed, capacitor C will start charging toward battery voltage E through resistor R. The time constant of RC will determine the rate of charge (or the slope of the curve). The current through C will become smaller and smaller as the capacitor is charged. At the same time, however, the voltage signal at point A is fed into an amplifier that has a very large voltage gain (A_E) and that also inverts the output at point B. This inverted signal is fed back to the bottom plate of capacitor C. The reason for the feedback is to increase the voltage difference across the capacitor plates, the top plate going positive and the bottom plate going negative. Charging current will now increase, producing a net result of capacitor C charging with a constant current.

Since the charge signal at A is then linear, the signal at point B is also linear. This point is the output (E_O), a negative-going ramp voltage.

In many applications where a Miller integrator is used, an *on/off* switch is required. The circuit of Fig. 5–22 shows a solid-state *on/off* switch found on many oscilloscope time bases.

When switch S is in position 2, both diodes CR_1 and CR_2 are forward-biased, and capacitor C is shorted. The voltage at point C (output) is a voltage equal to the voltage at point 2 ($+V$), less the drop across CR_2. In

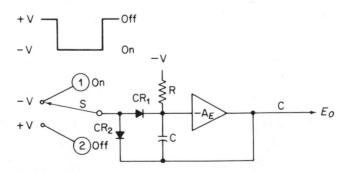

Figure 5–22. Miller integrator with *on/off* switch.

effect, CR_2 is a clamp diode, clamping the output to a known voltage level.

With switch S is in position 1, both CR_1 and CR_2 are reverse-biased. Point B starts charging toward a negative voltage. The output at point C is then a positive-going linear ramp (due to the inversion of the operational amplifier). If switch S is again set to position 2, capacitor C is discharged, and the output will return to the clamp voltage. This action is generally referred to as *flyback*. In effect, CR_1 is a switch diode or *on/off* diode.

In a practical solid-state circuit, switch S is replaced by two voltage levels, switching between $-V$ or *on* and $+V$ or *off*. Such an input waveform is often called the *gate* or *gate voltage*.

In any theoretical IC, the slope of the ramp is determined by the charging rate of capacitor C, in turn a function of the time constant, RC. In a practical IC, a variable timing resistor R and/or a variable timing capacitor C are used, if accurate adjustment of the ramp slope is required. If different slopes are needed, then a switching arrangement is used. Many times capacitors and/or resistors will be shared in the different ranges set by the switch.

In any practical circuit, the input impedance of the amplifier must be high so that the linearity of the integrator is not affected. When the amplifier is not in an IC package, the input stage is usually an FET connected in the source-follower configuration. This provides the high-input impedance. Typically, the remainder of the amplifier stages are direct-coupled transistors, arranged to provide high-gain inverted output with low-output impedance.

5–3.1. Failure Patterns for Integrators

The most common failure in an integrator, besides a complete breakdown, is nonlinearity of the ramp. If there is total breakdown, check the active elements in the usual manner. That is, measure waveforms and voltages at all stages of the amplifier, starting with an overall input/output measurement. This should pinpoint any major failure.

If the integrator is working, but the ramp is not linear, look for three major causes: a leaking capacitor C, a leaking gate diode CR_1, or low open-loop gain in the amplifier. If the RC circuit has more than one range, check all possible ranges for nonlinearity. One or a few nonlinear ranges indicate that the amplifier is probably good but that the capacitors and/or resistors may be defective.

If all ranges are nonlinear, check the amplifier gain. If convenient, check the amplifier open-loop gain against the service literature. If this is not practical or the open-loop gain is not specified, replace the feedback capacitor with a resistance, as described in Sec. 5–2.4.3. Check the amplifier for a closed-loop gain of at least 100 (feedback resistor 100 times that of the input capacitor). If the amplifier cannot produce a closed-loop gain of at least 100, it is usually unsatisfactory for use with an integrator.

If an amplifier shows low gain and has an FET at the input, look for a leaking FET. Experience has shown that the input FET most often is the active element which is defective.

5-4. BASIC DIFFERENTIAL AMPLIFIER

Differential amplifiers are used in many laboratory and industrial applications. A differential amplifier is an amplifier that uses the push/pull principle of operation and is generally used to amplify the *difference between two signals.*

Differential amplifiers are often found in IC form. Figure 5–23 is a block diagram of such an amplifier showing input and output signals. Notice that the signal at point A (E_A) is *with respect to signal ground,* and that the signal at point B (E_B) is also with respect to signal ground. The difference $E_A - E_B$ becomes the signal applied to the amplifier. The output will then be $E_A - E_B$ times the gain A_E. The output is taken between leads C and D and not from C to ground (E_C) or from D to ground (E_D). Terminal A or terminal B may be grounded, making the input to the amplifier a single-ended or grounded input. In this case, the circuit will become a single-ended input to balanced-output converter.

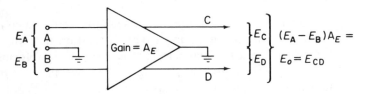

Figure 5–23. Block diagram of differential amplifier.

One of the main uses for a differential amplifier in laboratory work is that of an amplifier for meters, oscilloscopes, recorders, etc. Such instruments are operated in areas where many signals may be radiated (power-line radiation, stray signals from generators, etc.). Test leads connected to the input terminals will pick up these radiated signals, even when the leads are shielded. If a single-ended input is used, the undesired signals will be picked up and amplified. If the amplifier has a differential input, both leads will pick up the same signal at the same time. Since there is no difference between the signals at the two inputs, there will be no amplification. Thus, the undesired signals will.not appear on the output.

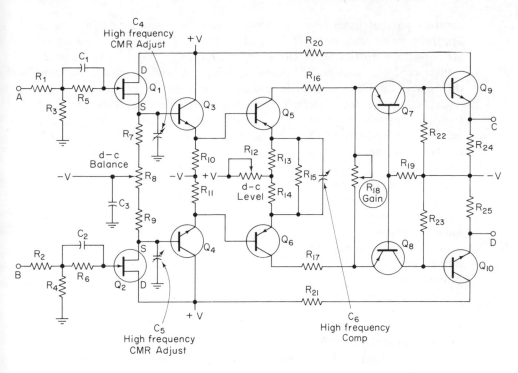

Figure 5–24. Differential amplifier used in laboratory instrument.

Figure 5–24 is the diagram of a differential amplifier used in a typical laboratory instrument. Note that FETs are used at the input (for high-input impedance) while the output uses the emitter-follower configuration (for low-output impedance).

In practical troubleshooting of any differential amplifier, other than looking for a major failure, the most important thing is that both halves of the amplifier be electrically symmetrical. That is the input impedance at A should be exactly equal to the input impedance at B. R_1 and R_3 form an input attenuator with an impedance of $R_1 + R_3$. The voltage attenuation is equal to the ratio $R_3/R_2 + R_4$. (Similarly, R_2 and R_4 form a division ratio of $R_4/R_2 + R_4$.) The sum of $R_1 + R_3$ is typically on the order of 1 mΩ. Resistors R_5 and R_6 (typically in the range of 100 K) limit the input current to protect Q_1 and Q_2. Capacitors C_1 and C_2 couple a-c signals directly to the gates of Q_1 and Q_2 and thus give high-frequency peaking.

Note that there are many adjustment controls in the circuit of Fig. 5–24. To troubleshoot such a circuit properly, you must know the effects of these controls on the circuit. For example, it may be possible to eliminate an apparent major fault simply by adjustment of controls. The following descriptions are applicable to the specific controls of Fig. 5–24. However, similar controls (possibly identified by different names) appear on most differential amplifiers used in laboratory instruments.

Capacitors C_4 and C_5 (high frequency) CMR adjust and shunt high-frequency signals to ground. Capacitor C_6 shunts high-frequency signals between the emitters of Q_5 and Q_6. These three controls form the high-frequency compensation adjustments for the amplifier. When properly adjusted, the high-frequency signals will be attenuated so that the response is flat over the desired range.

The d-c balance potentiometer R_8 compensates for any inherent differences between the two halves of the input circuit. It is impossible to match the two halves perfectly. Also, the component values will change with age and thus produce an unbalance. Generally, R_8 is a very temperature-stable potentiometer and mechanically secure so as not to lose its setting with any vibration. (Sometimes the potentiometer shaft is provided with a locknut to prevent a change in setting due to vibration.)

The functions of R_{12} (d-c level) and R_{18} (gain) are often confused. Potentiometer R_{12} sets the level of the d-c voltage at the output, while R_{18} sets the overall gain of the amplifier. It is quite possible for the amplifier to operate at the correct output level but not to provide the necessary gain, and vice versa.

In order to prevent temperature from causing the amplifier d-c levels to drift, transistors Q_1 through Q_4 are mounted on the same temperature heat sink. Sometimes, Q_1 through Q_4 are enclosed by a metal can, so they will all remain at the same temperature. In some cases, Q_5 and Q_6 may also share the same heat sink, but usually they are separate from Q_1 through Q_4. However, transistors Q_1 through Q_4 invariably are *matched pairs and should be replaced as such*.

Transistors Q_5 and Q_7 (as well as corresponding transistors Q_6 and Q_8) form a cascode amplifier, with the result that the input to Q_5 is voltage, while the output of Q_5 into Q_7 is current. The output of Q_7 is voltage again, since the input impedance to Q_7 is quite low, while the Q_7 output impedance is quite high.

5–4.1. Failure Patterns for Differential Amplifiers

The four most common failures of a differential amplifier are loss of gain (voltage or current), poor common-mode rejection ratio, d-c unbalance, and output signal d-c drift.

Loss of gain (voltage or current). If a loss of gain has been gradual, component aging is usually the fault. Always try to correct a loss-of-gain problem, frequency-response problem, or an improper output level problem by adjustment first. Gradual losses are not common in well-designed equipment. Also, usually only one-half of the circuit will lose gain, thus unbalancing the amplifier. Any differential amplifier that requires continued adjustment of the d-c balance control is suspect. If there is a total breakdown or extreme unbalance, check the active elements in the usual manner (waveform and voltage measurements at all stages, starting with overall input/output measurement). This should pinpoint any major failure.

Poor common-mode rejection. Before going into the problems of common-mode rejection (or CMR), we must discuss its definition. All manufacturers do not agree on the exact definition of common-mode rejection. One manufacturer defines common-mode rejection (CMR or CM_{rej}), or the common-mode rejection ratio (CMRR), as the ratio of differential gain (usually large) to common-mode gain (usually a fraction). That is, the amplifier may have a large gain of differential signals (different signals at input terminals) and a low gain (or a loss) of common-mode signals (the same signal at both terminals). Another manufacturer defines CMR as the relationship of change in output voltage to the change in the input common-mode voltage producing it, divided by the open-loop gain.

For example, assume that the common-mode input (applied to both terminals simultaneously) is 1 V, the resultant output is 1 mV, and the open-loop gain is 100. The CMR is then

$$\frac{0.001}{\frac{1}{100}} = 100,000 \text{ or } 100 \text{ dB}$$

Another method by which to calculate CMR is to divide the output signal by the open-loop gain to find an *equivalent differential input signal*. Then the common-mode input signal is divided by this equivalent differential input signal. Using the same figures as the previous CMR calculation,

$$\frac{1 \ mV}{100} = 0.00001 \text{ equivalent differential input signal}$$

$$\frac{1 \ V}{0.00001} = 100,000 \text{ or } 100 \text{ dB}$$

No matter what bassi is used for calculation, CMR is an indication of the degree of circuit balance of the differential stages, since a common-mode input signal should be amplified identically in both halves of the circuit. A large output for a given common-mode input is an indication of large unbalance or poor CMR. If there is an unbalance, a common-mode signal becomes a differential signal after the first stage.

As with amplifier gain, CMR usually decreases as frequency increases. However, as a rule of thumb, the CMR should be at least 20 dB greater than the open-loop gain at any frequency (within the limits of the amplifier).

It is obvious that poor CMR is due most commonly to nonsymmetrical gain on both halves of the amplifier. This is one of the reasons for including the CMR adjustment capacitors (C_4 and C_5 in Fig. 5–24).

Always try to correct a CMR problem by adjustment first. Do not forget to include the balance adjustments, since any unbalance will be reflected as a poor CMR ratio. Another point to remember is that attenuator probes, attenuator networks, and the like, must be perfectly matched, with regard to both a-c and d-c, to get a good CMRR. Try to avoid using external attenuators in calibrating a differential amplifier for CMR. Unmatched external calibrators just add one more unknown.

D-C unbalance. Problems of d-c unbalance are almost always due to component aging or mismatched temperature coefficients of symmetrical (that is, both halves) components. If many stages are used in the amplifier, start by balancing the *last stage first*. Make the inputs to the last stage equal (tie both halves together, ground both inputs, or apply the same signal to both inputs, depending on the specific circuit) and adjust the stage's d-c balance control. Sometimes, the last stage will not have a balance control but be checked for balance. For example, in Fig. 5–24, the voltage drop across R_{20} should be the same as across R_{21}, the drop across R_{24} should be the same as R_{25}, and so on.

Continue this process, moving toward the first stage and always monitoring *the main output,* until the trouble has been located. Usually, this sequence can isolate a badly unbalanced section. At this point, troubleshoot the suspected stage by voltage measurements in the usual manner.

If the circuit is part of an oscilloscope (differential amplifiers are often found in laboratory oscilloscopes), an unbalance is easily detected by grounding the two inputs and rotating the oscilloscope gain control (often called the venier control). An unbalance will show up in the form of a trace shift. That is, the oscilloscope trace will shift vertically as the gain control is changed, even though both inputs are at zero. This is a result of a d-c signal being developed across the gain control due to unbalance. The amplifier can then be balanced by adjusting the amplifier d-c balance potentiometer until no trace shift is observed when the gain control is rotated.

Output signal drift. A slow drift of any kind at the output is usually linked to a d-c unbalance. Temperature compensation (or a lack thereof) plays a major role in such cases. A well-designed circuit is very seldom plagued by temperature problems. However, occasionally circuits may be overheated, and solid-state circuits can be partially damaged.

When replacing parts in a differential amplifier, always make sure that both halves of the amplifier are well matched, especially in the very first stages.

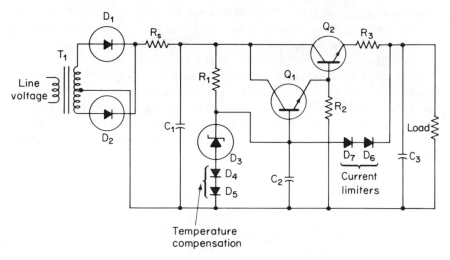

Figure 5–25. Typical nonfeedback power supply.

5–5. BASIC POWER SUPPLIES

There are two basic types of regulated solid-state power supplies: those with feedback and those without.

Figure 5–25 is the schematic of a typical nonfeedback supply. As shown, D_1, D_2, and C_1 form a full-wave rectifier. R_s is a small-valued resistor to limit surge current. R_1 and Zener diode D_0 form a reference circuit. Capacitor C_0 reduces the ripple and Zener noise. Regulation is accomplished by Q_1 and Q_2. If the output voltage increases, the base of Q_1 remains constant (due to the action of D_3) but the current through Q_1 increases (due to the increase in Q_1 collector voltage). This increases the emitter voltage of Q_1, and thus the base voltage of Q_2. The increase in Q_2 base voltage causes an increase in Q_2 current. The increase in output voltage is thus offset. Diode D_3 has a positive temperature coefficient. D_4 and D_5 have negative temperature coefficient. Thus, by combining these two properties, the output voltage can be made relatively independent of temperature changes.

Diodes D_6 and D_7 are silicon diodes that start conducting heavily at approximately 0.6 V (as opposed to 0.2 V for the junctions of Q_1 and Q_2, both germanium transistors). Under normal load conditions, D_6 and D_7 are forward-biased but nonconducting. Under typical conditions there are 0.2-V drops across Q_1 and Q_2, as well as a 0.3-V drop across R_3. This provides a total drop of 0.7 V. A voltage of 1.2 is required for D_6 and D_7 to conduct (0.6 V + 0.6 V = 1.2 V).

If the load is accidentally shorted, the drop across R_3 goes up. When the drop across R_3 equals 0.8 V, D_6 and D_7 begin to conduct heavily. As a result, biasing current for Q_2 is shunted through D_6 and D_7. This limits the amount of current that can be passed by Q_2. Diodes D_6 and D_7 thus provide an effective method of current limiting for the supply.

Figure 5–26 is the schematic of a typical feedback supply. As shown, D_1, D_2, and C_1 form a full-wave rectifier. The function of reference Zener diode D_6, current limiting diodes D_3 through D_5, and the control transistors Q_1 and Q_2 are essentially the same as for corresponding parts on the nonfeedback supply (Fig. 5–25). However, transistors Q_3 and Q_4 form a feedback circuit to sense variations in the output voltage.

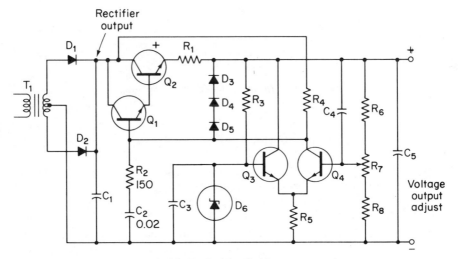

Figure 5–26. Typical feedback power supply.

The voltage at the base of Q_4 depends on the setting of R_7 (which sets the power supply voltage, within limits) and the output voltage. Any variation in output voltage causes the base of Q_4 to vary. The base of Q_4 is held constant, since it is coupled to the emitter of Q_3 (which in turn is held constant by the Zener diode D_6 voltage at the base). In effect, Q_3 is an emitter-follower (with constant emitter and base), while Q_4 is a gain stage. The output of Q_4 is applied to the base of Q_1, thus providing the feedback regulator action.

For example, assume that the output voltage goes more positive (due to a change in the load or line voltage). This makes the base of Q_4 more positive, the collector of Q_4 and base of Q_1 more negative, and the emitter of Q_2 more negative, thus offsetting the original change.

Resistor R_2 and capacitor C_2 form a phase shift or lag network to prevent oscillation in the feedback system. Capacitor C_4 (often called a *speed-up* capacitor) passes any ripple or noise directly to the base of Q_4, thus providing effective control of these undesired changes in output voltage. Ripple, noise, and so on are seen as output variations by a feedback regulator and therefore are suppressed (in a properly working circuit).

5–5.1. Failure Patterns for Power Supplies

The obvious failure of any power supply is failure to deliver the correct voltage to a given load. The failure of a regulated supply is failure to deliver the rated voltage in the presence of changes in line voltage or load (within limits) and failure to suppress ripple, noise, and the like (within limits). Therefore, the obvious first test of a power supply is to monitor the output voltage. If convenient, the next test is to vary the load and the line voltage over their rated limits and see if the output voltage remains constant (within limits). The final test is to monitor the output voltage with an oscilloscope, checking for ripple, noise, and the like.

If the output voltage remains within limits when the line and load are varied and the ripple is within tolerance, the power supply is operating properly. If not, an analysis of the symptoms can sometimes pinpoint the problem without further measurements.

One technique often overlooked when troubleshooting solid-state power supplies is monitoring the ripple voltage at the rectifier output and regulator (or final) output. An analysis of the ripple voltage and frequency can sometimes pinpoint trouble.

For example, the presence of ripple voltage at the rectifier output (D_1 and D_2 common) is normal. (Even a large value of C_1 cannot prevent ripple.) A 1-V ripple signal at the rectifier output is not unusual for a supply with a 12-V output (typical solid-state supply). The presence of ripple at the regulator output (power supply output) is not normal, even with the nonfeedback supply. Typically, a feedback supply will have a ripple less than 10 mV (often 1 mV or less in a good circuit), while a nonfeedback regulator will have difficulty in maintaining a ripple of less than 10 mV (unless C_1 is very large).

The ripple frequency is twice the line frequency in a full-wave rectifier. Thus, with a 60-Hz line, the ripple at the output of D_1 and D_2 should be 120 Hz. If the ripple is 60 Hz at the rectifier output, this indicates that D_1 or D_2 is defective (open). If the ripple pattern is not symmetrical on the oscilloscope display (one cycle is smaller and possibly distorted from the other), this indicates D_1 or D_2 is partially defective (possibly leaking or partially shorted).

There should be no ripple at the regulator output. However, there may be some line frequency hum or noise (due to stray pickup, ground loops, etc.) present at any point in the circuit. However, such hum or noise will be at the line frequency, or 60 Hz. If the signal at the output is 120 Hz, this definitely indicates that the regulator is not suppressing the rectifier ripple.

If trouble is traced to a regulator, the problem can be traced through each stage by the usual methods: shorting base to emitter for *turn off* and connecting collector to base through a resistance for *turn on*. (Refer to Chapter 2.) Feedback power supplies are relatively easy to troubleshoot since they are direct coupled.

One possible exception to shorting emitter to base for turn off is in a power transistor that may not turn off immediately, even though it is not defective. Such transistors will turn off in time, after they have cooled down. However, they may give the appearance of not turning off. The following paragraphs summarize the effects on the power supply of component failure.

Failure of either D_1 or D_2 will cause a change of ripple frequency and a reduction of voltage. If C_1 is shorted, D_1, D_2, and possibly the transformer will be burned out. If C_1 is open, the ripple will increase and regulation will be poor. A leaky C_1 can both reduce voltage and produce poor regulation. The value of C_1 must be large enough to keep ripple low but not too large so as to exceed the surge ratings of D_1 and D_2.

If D_3 in Fig. 5–25 (or D_6 in Fig. 5–26) is shorted, the output voltage will drop to a low value (usually below 1 V). If D_4 or D_5 in Fig. 5–25 is shorted, there will be only a slight drop in output voltage (about 0.6 V for each diode). However, regulation will become poor, especially in the presence of temperature change (such as a prolonged overload where all of the transistors are heated).

If R_1 is shorted, D_3 will probably be destroyed. If R_1 opens, there will be no regulation. If R_1 decreases in value, there will be little effect (except in the case of a complete short). However, if R_1 increases in value, D_3 may operate near the breakdown point and become noisy. Zener diodes can produce noise (erratic voltage variation) if operated too near their breakdown voltage.

If C_2 in Fig. 5–25 shorts, there will be no regulation and R_1 may burn out. A leaky C_2 can produce poor regulation. An open C_2 usually shows up as excessive ripple or noise. Capacitor C_2 must be large enough to be an effective short to ripple frequencies, but not so large as to make C_2 and Zener D_3 a relaxation oscillator. (This does not occur as often in solid-state Zener diodes as it does with gas-tube regulators.)

If either or both D_6 and D_7 in Fig. 5–25 are shorted or badly leaking, there will be a reduction in output current. If heavy current is being drawn, voltage will drop. If D_6 or D_7 open, there may be no immediate effect, except that the power supply will lose its current overload protection.

If C_3 in Fig. 5–25 (or C_5 in Fig. 5–26) is shorted, any number of components can be burned out, even with the overload protection of D_6 and D_7. If C_3 or C_5 open, there may be no pronounced effect on the power supply operation. Capacitors C_3 and C_5 often provide very little regulator or filter action. Their main purpose is to provide a low impedance path for signals in the circuit being supplied power. Badly leaking C_3 or C_5 capacitors usually result in low-output voltage.

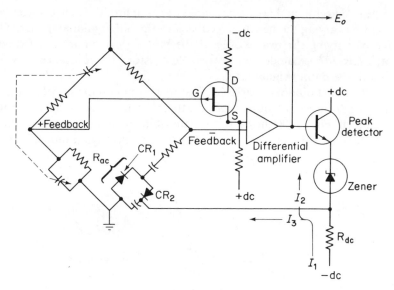

Figure 5–27. Solid-state Wien bridge oscillator.

5–6. WIEN BRIDGE CIRCUITS

Variations of the basic Wien bridge circuit are often used in laboratory equipment. A solid-state Wien bridge oscillator is a typical example. The basic circuit is shown in Fig. 5–27.

The circuit consists essentially of a Wien bridge (containing both resistive and reactive components), a differential amplifier, and a peak detector. To use a Wien bridge as an oscillator, the bridge output is amplified, shifted in phase by 360°, and returned to the bridge as an input. In theory, there will be an exact 360° phase shift and sufficient feedback only at the resonant frequency of the Wien bridge (where the reactive component equals the resistive component and impedance is zero). In practice, however, the bridge is operated slightly off balance to produce a small output signal, since bridge output at true resonance is zero.

In the circuit of Fig. 5–27, the positive feedback (reactive output) exceeds the negative feedback (resistive output). The positive feedback is applied to an FET which acts as a buffer to match the bridge and differential amplifier. The negative feedback is also applied to the differential amplifier. The difference (positive feedback less negative feedback) is amplified and fed to the output, as well as used to drive the bridge.

The main advantage of a Wien bridge oscillator is stability. The Wien bridge oscillator offers the greatest amount of stability for a variable-frequency oscillator. To stabilize the output in Fig. 5–27, the output is sampled by the peak detector to detect any changes in amplitude. Such changes are corrected by varying the negative feedback to the differential amplifier. This is done by varying the resistive leg of the bridge through action of diodes CR_1 and CR_2. These diodes act as variable resistors, the resistance of which varies with the output signal level.

Since the use of diodes as variable resistors is common in many laboratory circuits, it is helpful in troubleshooting to understand the basic technique. The graph of Fig. 5–28 shows the relationship of diode voltage to diode current. Assume a d-c current of 2 mA is flowing through a diode. The a-c resistance at this operating point is

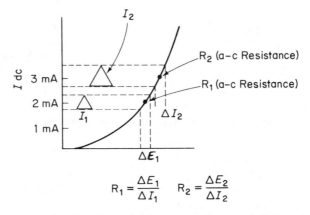

$$R_1 = \frac{\Delta E_1}{\Delta I_1} \qquad R_2 = \frac{\Delta E_2}{\Delta I_2}$$

Figure 5–28. Graph showing how a-c resistance of diodes can be varied by dc.

$$R_1 = \frac{\text{diff } E_1}{\text{diff } I_1}$$

Assume that the d-c current is increased to 3 mA. The operating point is then shifted up the curve and the a-c resistance is

$$R_2 = \frac{\text{diff } E_2}{\text{diff } I_2}$$

Since the curve is steeper at this point, the same small change in voltage produces a greater change in current (diff I_2). Thus the a-c resistance is lower at this point. If the d-c current is decreased to 1 mA, the a-c resistance will increase. Thus, the diode acts as a variable a-c resistor controlled by d-c.

In the circuit of Fig. 5–27, the diode control current flows from the minus supply through resistor R_{DC} to ground through the forward-biased diodes CR_1 and CR_2. This same current also flows through the peak detector transistor Q_1. Thus, the current I_1 is divided into two currents I_2 and I_3. The by voltage drop across R_{DC} is held constant (almost) by the drop across the diodes and the Zener. Thus, the current I_1 remains fairly constant, and if the I_2 current increases, the I_3 current will decrease ($I_3 = I_1 - I_2$). The I_2 current depends on the amplitude of the output signal. Therefore, the resistance produced by the diodes is controlled directly by the output amplitude.

As an example, assume that the output increases. This causes an increase in I_2 current and a corresponding decrease in I_3 current. The lower I_3 current causes an increase in the a-c resistance of the diodes, thus increasing the negative feedback and reducing the output.

5–6.1. Failure Patterns for Wien Bridge Circuits

In much equipment, the Wien bridge portion of the circuit is provided with several ranges. The ranges are selected by switches which connect resistors and capacitors in and out of the circuit to cover a wide frequency range. If this is the case, the first step is to check the operation of the equipment on all ranges. If a particular problem (instability, distorted waveform, low amplitude, off-frequency operation) appears only on one or a few ranges, check the components in the switching and bridge circuits, particularly those components in the affected ranges.

If the problem appears on all ranges, check the active elements (transistors, IC package amplifier, etc.) in the usual manner (waveform and voltage measurements). If this does not pinpoint the problem, a study of the following notes may help.

Instability (output amplitude varies) is usually the result of a fault in the stabilizing circuit (peak detector, control diodes, etc.). However, instability can also be caused by a problem in the amplifier. If possible, check the differential amplifier separately (without the feedback loop) as described in Sec. 5–4.

If the output is clipped or otherwise distorted, the problem is almost always in the amplifier. Look for transistor leakage, a shift in bias point, or possibly coupling capacitor leakage.

If the output appears to be undistorted and stable but low in amplitude, check the amplifier for a gain control. Try correcting the problem by adjustment. If the gain problem appears only on the low-frequency ranges or is worse at low-frequencies, look for a bad FET.

6. TROUBLESHOOTING TV RECEIVERS

This chapter is concerned with practical troubleshooting techniques for solid-state TV receivers. It is assumed that the reader is already familiar with vacuum-tube TV troubleshooting procedures for both black and white and color. Therefore, there will be no detailed discussion concerning use of basic TV test equipment, basic TV receiver theory, or the purpose of individual stages in a TV receiver, since these subjects are common to both vacuum-tube and solid-state television.

However, since it is necessary to understand the operation of a circuit to do a good troubleshooting job, a brief discussion of operational theory for each solid-state circuit is given.

A consistent format is used throughout the chapter for each circuit. First, a typical solid-state TV circuit diagram is given, along with the basic theory of operation. Second, the recommended troubleshooting approach for that particular type of circuit is discussed. Then a group of typical troubles is described, along with the most likely causes of such troubles.

6–1. LOW-VOLTAGE POWER SUPPLY

Figure 6–1 is the schematic diagram of a typical low-voltage power supply for solid-state TV receivers. The circuit consists essentially of a full-wave bridge rectifier, followed by a solid-state (Zener diode and transistor) regulator. As with many solid-state TV receivers, the set can also be operated with self-contained rechargeable batteries; therefore, the power supply circuit also provides for recharging the batteries. The set can also

238

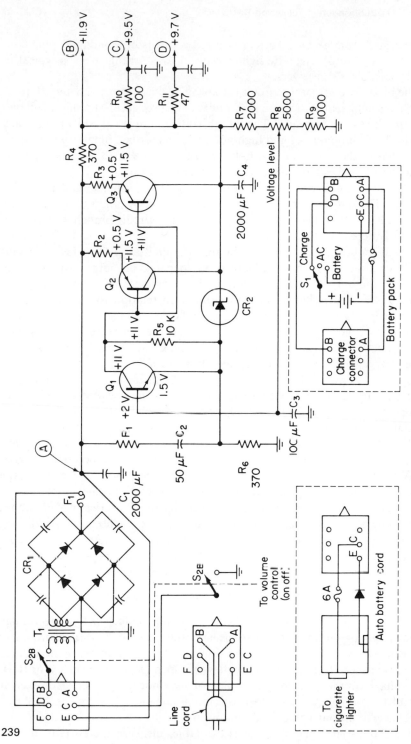

Figure 6–1. Low-voltage power supply.

239

be operated from an auto battery (12 V) by means of an adapter plug connected to the auto's cigarette lighter output. This arrangement is typical of small portable TV receivers.

The basic function of the circuit is to provide d-c at approximately 10 to 12 V. This is sufficient to operate all circuits of the set except the high voltages required by the picture tube. These high voltages are supplied by the flyback circuit (horizontal output and high voltage) as described in later paragraphs. Note that in the circuit of Fig. 6–1 the negative side of the power supply is grounded. Positive grounding is used equally as often.

The line voltage is dropped to about 12 V by transformer T_1 and is rectified by CR_1. The output from CR_1 is regulated by Q_1, Q_2, and Q_3, as well as by Zener CR_2, and is distributed to three separate circuit branches, each at a slightly different voltage level.

Operation of the regulator is standard. The emitter of Q_1 is held constant by Zener diode CR_2 while the base of Q_1 depends on output voltage. Any variation in output voltage (resulting from changes in input voltage or variations in load) change the base voltage in relation to emitter voltage. These changes appear as a variation in Q_1 collector voltage and, consequently, $Q_2 - Q_3$ base voltage. Transistors Q_2 and Q_3 are connected in series with the output of CR_1 and the load. Thus, Q_2 and Q_3 act as variable resistors to offset any changes in output voltage. For example, if the output voltage increases, the base of Q_1 swings more positive, causing a drop in Q_1 collector voltage. This causes the bases of Q_2 and Q_3 to swing negative and results in more collector current flow (since Q_2 and Q_3 are PNP). Thus, the increase in output voltage is offset. The level of the output voltage is set by R_8.

When the set is to be operated on battery power, switch S_1 is set to *battery*, the a-c power plug is removed and the battery pack cord is plugged in. When the battery is to be recharged, switch S_1 is set to *charge* and the a-c power cord is plugged into a special connector on the battery pack. Although there is no standardization, rechargeable batteries provide about 4 to 6 hours of operation and require about 8 to 12 hours of recharging time. When the set is to be operated from an auto's cigarette lighter, the a-c power plug is removed and the auto battery power cord is plugged in.

6–1.1. Recommended Troubleshooting Approach

If the symptoms indicate a possible defect in the low-voltage supply, the obvious approach is to measure the d-c voltage. If there are many branches (three output branches are shown in Fig. 6–1), measure the voltage on each branch. If any of the branches are open (say, an open R_{10} or R_{11}), the voltage on the other branches may or may not be affected. However, if any of the branches are shorted, the remaining branches will probably be affected (the output voltage will be lowered).

If a clamp-type current probe is available, measure the current in each branch. This is not always possible. Many solid-state sets use printed circuit boards where the output leads are etched wiring rather than a wire or group of wires.

The use of printed circuit boards also eliminates the possibility of disconnecting each output branch, in turn, until a short or other defect is found. This practice is common on vacuum-tube sets or those without printed wiring.

If one or more output voltages appears to be abnormal and it is not practical to measure the corresponding current, remove the power and measure resistance in each branch. Compare actual resistances against those in the service literature, if available. If you have no idea as to the correct resistance, look for obvious low resistance (a complete short or a resistance of only a few ohms).

Always use an isolation transformer for the a-c power line. In many solid-state circuits, one side of the a-c line is connected to the chassis, even though the set may have a power transformer. As a convenience, have a 12-V d-c supply available to substitute for the low-voltage power supply circuit of the set. A 12-V battery eliminator makes a good source if a conventional power supply is not available. Ideally, the substitute power supply should be adjustable (in output voltage) and have a voltmeter and ammeter to monitor the output.

If the set can be operated on batteries, switch to battery operation and see if the trouble is cleared. If so, the problem is definitely localized to the low-voltage power supply circuit. Likewise, if the set can be operated from an auto battery or similar arrangement, switch to that mode and check operation.

If it becomes necessary to replace a Zener diode in the regulator circuit, always use an exact replacement. Some technicians will replace a Zener with a slightly different voltage value and then attempt to compensate by adjustment of the regulator circuit. This may *or may not* work.

Do not overlook the use of an oscilloscope in troubleshooting the power supply circuits. Even though you are dealing with d-c voltages, there is always some ripple present. The ripple frequency and the waveform produced by the ripple can help in localizing possible troubles. Also an oscilloscope with a d-c/a-c vertical input switch will work as a d-c voltmeter.

6–1.2. Typical Troubles

The following paragraphs discuss symptoms that could be caused by defects in the low-voltage power circuits.

6–1.2.1. No Sound and No Picture Raster

When there is no raster on the picture-tube screen and no sound whatsoever, it is likely that the low-voltage power circuits are totally inoperative or are producing a very low voltage (less than 20 to 30 per cent). If the circuits are producing a voltage about 50 per cent of normal (say, 6 V in a 12-V system), there will be some sound, even though the raster may be absent. (This symptom is discussed in a later section.) The most likely defects are filter capacitor leakage, a defect in the regulator (regulator completely cut off), or a short in the output line.

Start the troubleshooting process by making voltage measurements at test points A, B, C, and D, or their equivalent.

If the voltage is absent at all test points, check all fuses and switches as well as CR_1 and T_1. If the voltage at B, C, and D is absent or very low but the voltage at A is high, the regulator circuit is probably at fault. If the voltage is low at A, the filter capacitor C_1 is probably at fault. If all the voltages are low, look for a short in one or more of the output lines.

To check the regulator, first make sure that all three transistors are forward-biased. Typically, the base of Q_1 will be about 2 V, with an emitter voltage of 1.5 to 1.8 V. The bases of Q_2 and Q_3 will be about 11 V, with the emitters at 11.5 V. In any event, all three transistors must be forward-biased for the regulator to operate.

If any of the transistors are not forward-biased, remove power and check the transistors in circuit (with an in-circuit tester) or out of circuit (ohmmeter method) as described in Chapter 2. If the transistors appear to be good but not forward-biased, check all of the associated resistors and capacitors with an ohmmeter on a point-to-point basis.

Note that it is possible for transistor Q_1 to be cut off while Q_2 and Q_3 are forward-biased. However, such a condition will quickly point to a fault in Q_1 or its associated parts.

A common fault in such regulator circuits is a base-emitter short in the current-carrying transistors Q_2 or Q_3. Since these transistors are in parallel, it may be difficult to tell which transistor is at fault. If necessary, disconnect each transistor and check it separately. If either Q_2 or Q_3 has a base-emitter short, the regulator will be cut off, and both the base and emitter will be high.

6–1.2.2. No Sound, No Picture Raster, and Transformer Buzzing

These symptoms are similar to those described in Sec. 6–1.2.1. except that the transformer is buzzing or humming. Such a condition indicates excessive current. The most likely defects are a short circuit ahead of the regulator (test point A, for example), shorted rectifiers, or shorted turns in the power transformer.

Again, start the troubleshooting process by making voltage measurements at test points A, B, C, and D. Keep in mind that prolonged excessive current will cause one or more fuses to blow. Therefore, with these symptoms the assumption must be excessive current that is below the rating of the fuses. Such a condition will normally result in very low voltage, but voltage will not be completely absent.

Unlike the previous symptoms, the regulator is probably operating properly within its capabilities. Therefore, concentrate on short circuits, particularly the rectifiers, and possibly shorted transformer turns. Both T_1 and CR_1 can be checked by substitution or by an ohmmeter test, whichever is convenient.

If this does not localize the problem, check for short circuits in each branch of the output. Also look for overheated parts, such as the transformer or burned printed circuit wiring.

6–1.2.3. Distorted Sound and No Raster

These symptoms are similar to those described in Sec. 6–1.2.1, except that there is some sound (usually with intercarrier buzz). Generally, this indicates that the power supply output voltage is below normal but not absent or near zero. These symptoms are produced when the power supply

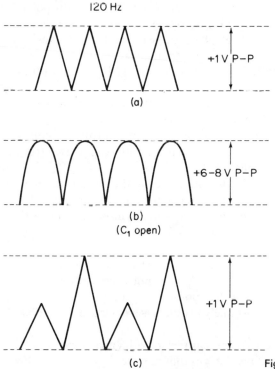

120 Hz

+1 V P–P

(a)

+6–8 V P–P

(b)
(C_1 open)

+1 V P–P

(c)
(CR_1 partially defective)

Figure 6–2. Typical solid-state, regulated power-supply waveforms.

output is about 40 to 60 per cent of normal. The most likely defects are the filter capacitor, rectifiers, and power transformer.

Start the troubleshooting process by making voltage measurements at test points A, B, C, and D or their equivalents. Next, measure the amplitude and frequency of the waveforms at the same test points, particularly at the bridge output (test point A) and the regulated d-c output (test point B). An analysis of the waveforms in a solid-state supply can often pinpoint troubles immediately. The following are some examples.

The normal waveform at the bridge output (test point A) is a sawtooth (almost) at twice the line frequency (usually 120 Hz), as shown in Fig. 6–2 a. If the filter and regulator are operating properly, the 120-Hz signal will be suppressed at the final output. However, there may be some line frequency (60-Hz) ripple at the regulator output (all three branches).

If there is a strong 120-Hz waveform at the regulator output, this indicates that the filter and/or regulator are not suppressing the 120–Hz bridge output. The most likely cause is an open filter capacitor C_1 or a regulator circuit defect. Capacitor C_1 could be leaking. However, excessive leakage or a short would blow the fuse.

Capacitor C_1 can be checked for an open by connecting a known good capacitor in parallel. Make sure to use the right capacity and right polarity (if C_1 is an electrolytic). Also avoid connecting the test capacitor leads across C_1 with the power applied. The voltage surge could damage transistors throughout the set. Remove the power, connect the test capacitor, and then reapply the power. A test capacitor will not show leakage in C_1. Substitution is the only true test.

The waveform at test point A will also indicate the condition of capacitor C_1. If the waveform changes from an almost sawtooth 120 Hz to a 120-Hz half sinewave (Fig. 6–2b), capacitor C_1 is probably open. This is confirmed further if the amplitude of the waveform at test point A increases from a typical 1 V (or less) to several volts.

The condition of rectifier CR_1 is also indicated by the waveform at A. If the waveform is 60 Hz (line frequency) then one-half of rectifier CR_1 is defective (most likely shorted or open). If the waveform is not symmetrical (Fig. 6–2 c), one diode in CR_1 is probably defective (most likely leaking or having a poor front-to-back ratio).

Note that if any of the diodes in CR_1 are shorted, transformer T_1 will probably run very hot. Therefore, a hot T_1 does not necessarily indicate a bad T_1. However, if the waveform at test point A is good (indicating a good CR_1 and C_1) but T_1 is hot, the most likely problem is a defective T_1 (probably shorted turns).

Note that all four diodes in CR_1 are paralleled with capacitors to protect the diodes in case of sudden voltage changes that might exceed the breakdown voltage. If one of these capacitors is shorted, this can give the appearance of a shorted diode. If CR_1 is a sealed package with self-contained capacitors, the entire package must be replaced. If the capacitors can be replaced separately from the diodes, make sure that *both* the capacitor and diode are good before replacing either. For example, if the capacitor is open, the corresponding diode may be damaged. If the diode is replaced, the trouble will be repeated unless the capacitor is also replaced.

Keep in mind that an increase in load (say, due to a short or partial short) will cause an increase in ripple amplitude, even with a good filter and regulator. This is because a larger load (lower load resistance) causes faster discharge of the filter capacitor between peaks of the sinewave pulses from the rectifier.

6–1.2.4. Picture Pulling and Excessive Vertical Height

Thus far, we have discussed symptoms and troubles that result from a low power supply output due to defective parts or shorts. It is possible that the power supply can produce a high output voltage, resulting in picture pulling (the raster is stretched vertically and is bent). Hum bars across the picture screen are usually present with this condition.

Assuming that the trouble is definitely in the low-voltage supply, the most likely cause is in the regulator circuit. A defect in the rectifier, filter, or transformer is likely to result in a *low-voltage* output, producing the symptoms described previously.

It is possible for the regulator circuit to be cut off, causing the output lines to increase in voltage. The logical approach is to start the troubleshooting process by making voltage and waveform measurements at all test points.

If the voltage at test point B is nearly the same as at test point A (within about 0.5 V), the regulator is probably cut off. The voltage at test points C and D will also be high. The condition can be confirmed further if the waveform at test points B, C, and D is at 120 Hz (indicating that the regulator is not suppressing the 120-Hz bridge output).

With the trouble definitely pinned down to the regulator, make the forward-bias tests described in Sec. 6–1.2.1. This will localize trouble within the regulator circuit.

6–2. HIGH-VOLTAGE SUPPLY AND HORIZONTAL OUTPUT

There is very little standardization in the high-voltage and horizontal output circuits of solid-state TV receivers. There two basic circuit types: the *hybrid* circuit, which uses a vacuum tube (usually subminiature) for the high-voltage rectifier, and the *all-solid-state,* which uses solid-state rectifiers. At one time, most sets were of the hybrid type. Later, the hybrid

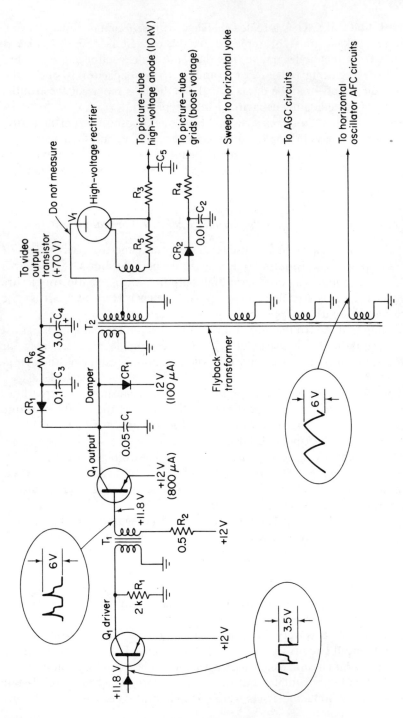

Figure 6–3. High-voltage supply and horizontal output circuit (hybrid type).

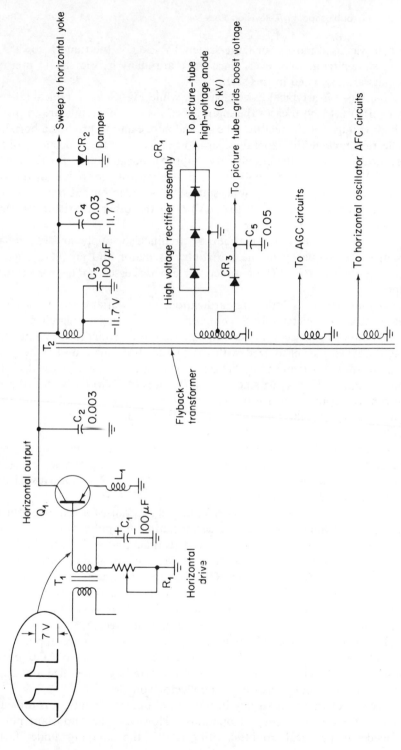

Figure 6–4. High-voltage supply and horizontal output circuit (all-solid-state type).

circuit was used mostly for large-screen TV. Today, the trend is toward all-solid-state circuits. However, since there are many hybrid sets in use, both circuits are discussed in this section.

Figure 6–3 is a typical hybrid circuit, while Fig. 6–4 is a typical all-solid-state circuit. In both cases the main functions of the circuits are to provide a high voltage for the picture-tube second anode and to provide a horizontal deflection voltage (horizontal sweep) to the picture-tube deflection yoke. In practically all cases the circuits also supply a boost voltage for the picture-tube focus and accelerating grids and possibly a voltage source for the video output transistor. (In many solid-state sets the video output transistor operates at a higher voltage (40 to 70 V) than the other transistors (typically 12 V).

In some cases the horizontal output circuits also supply an AFC signal to control the frequency of the horizontal oscillator and an AGC signal to provide *keyed AGC*. These functions are discussed in later sections as applicable.

Although there is little standardization, most horizontal output circuits have certain characteristics in common that must be considered during the troubleshooting process. The circuit receives pulses from the horizontal driver (at 15,750 Hz) synchronized with the picture transmission. The horizontal output transistor is normally biased at or near zero so that one edge of the pulse (negative in this case since the transistor is PNP) will drive the transistor into heavy conduction (near saturation), while the opposite swing of the pulse will cut off the horizontal output transistor. In effect, the horizontal output transistor is operated in the switching mode.

If the secondary winding of the driver transformer has some winding resistance, there will be some reverse bias (the average value of the base current pulses). This reverse bias could be as high as 2 V, but is usually less. However, the d-c bias will be near zero.

The collector current of the horizontal output transistor is applied through a winding on the flyback transformer, resulting in a pulse waveform at all of the other windings. The high voltage is rectified and applied to the picture-tube anode, while the boost output is rectified and applied to the picture-tube focus and accelerator grid anode. On some sets a separate winding is provided for the horizontal deflection yoke. On other sets the horizontal output transistor current is passed through the deflection yoke. Either way the horizontal output pulses produce the horizontal sweep.

The AFC and AGC windings (not found on all sets) are applied to the horizontal oscillator AFC circuits, and the tuner/IF stage AGC circuits, respectively. In some sets the output of one winding is rectified and applied to the video output transistor as a collector voltage.

Because of the great variety in horizontal output circuits, it is difficult to arrive at a typical theory of operation. However, the important point to consider in practical troubleshooting is that the damping diode starts to conduct when the transistor is cut off. The diode continues to conduct until the horizontal output transistor conducts.

During the horizontal forward scan (when the picture is displayed), the diode conducts and the transistor is cut off from the start of the sweep to about the midpoint. Then the diode is cut off, and the transistor conducts for the remaining half of the sweep. This sequence is shown in Fig. 6–5.

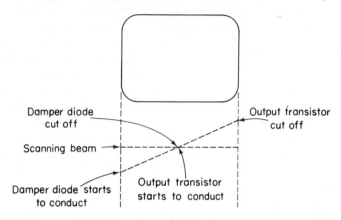

Figure 6–5. Relationship of damper diode and output transistor conduction periods to picture display horizontal sweep.

The scan sequence is important in troubleshooting, since any problems in the right-hand side of the picture are probably the result of defects in the transistor (or related components), while trouble in the left-hand side is probably the result of a defective damping diode.

When the transistor is cut off, the picture tube is blanked and the current flows rapidly through the horizontal yoke in the opposite direction, pulling the electron beam back to the left side of the screen (known as horizontal retrace or flyback). It is the pulse developed during the flyback interval that is used by the other windings on the flyback transformer.

Keep in mind that voltages in solid-state circuits are much lower than in corresponding vacuum-tube circuits. However, the currents are much larger in solid-state. For example, the emitter-collector current in some horizontal output transistors is almost 1 A (800 to 900 mA is typical). It is not practical to measure such current, except possibly with a clamp-type probe. Also resistances in solid-state coil windings are generally lower than in vacuum-tube circuits. Often, the horizontal deflection yoke winding is a fraction of 1 Ω.

6–2.1. Recommended Troubleshooting Approach

Many trouble symptoms caused by horizontal output circuits can also be caused by defects in other circuits. A dark screen (no raster) or insufficient width are two good examples. If the low-voltage power supply is completely inoperative (or almost zero), the screen can be dark. If the low-voltage supply is producing a low output, the picture width can be decreased. Of course, in either case the sound is absent or abnormal. On the other hand, if the horizontal oscillator and drive circuits are defective (no drive signal to the horizontal output tube), there will be no high voltage or horizontal sweep. Do not overlook the fact that the picture tube might be defective.

The most logical troubleshooting approach is to analyze the symptoms and then isolate the trouble to the horizontal output circuits by an input waveform check. That is, if sound is normal (indicating a good low-voltage supply), measure the input waveform (from the horizontal drive circuit) at the horizontal output transistor base. Generally, this is on the order of 6 to 8 V and appears similar to that shown in Figs. 6–3 and 6–4. Check the waveform against the service literature.

If the input waveform is normal, then trouble is in the horizontal output circuit (unless the picture tube is bad). Of course, if the input signal is not normal, the next step is to check the horizontal drive circuits as described in later paragraphs.

In the case of a completely dark screen, the next obvious check is to measure the high voltage at the picture-tube second anode. Then measure the accelerating grid and focus voltages from the boost circuit. If the voltages are present, the picture tube is defective. (Of course, check that the picture-tube filament is lighted.)

If either the high voltage or the boost voltage is absent or abnormal, this will isolate the trouble to the corresponding circuit. If both voltages are absent (with the drive signal good), the problem is in the horizontal output transistor or related circuit parts (capacitors, flyback transformer, etc.)

If only the high voltage is absent, check for a-c at the anode side of the high-voltage rectifier (unless this is not recommended on the service literature schematic). When measuring the high voltage, always use a meter with a high-voltage probe. Observe all of the usual precautions when measuring high voltage. In addition, *do not* make an arc test of the high voltage with solid-state circuits. The transient voltages can damage the transistors.

Measure the boost voltage with a meter and low-capacity probe. With the exception of the high voltage, most of the voltages can be measured using a low-capacity probe and a meter or oscilloscope. This is because the voltages are generally lower than in vacuum-tube circuits. Five hundred volts peak to peak is usually maximum for any solid-state circuit. Observe any precautions on the service literature schematic. Typically, there can be warning notes on the schematic such as "Do Not Measure High a-c Voltages."

It is often helpful if you can measure the emitter-collector current of the horizontal output transistor. Usually this is not practical, so you must use d-c voltage measurement to supplement the waveform checks. If the d-c voltages and waveforms at the horizontal output transistor are correct, it is safe to assume that the circuit is good up to the flyback transformer.

Do not use the same flyback checkers, yoke testers, horizontal system checkers, and the like used in vacuum-tube sets. Solid-state flyback transformers have different characteristics (Q, impedance, etc.) than vacuum-tube transformers. Likewise, the drive voltages in solid-state circuits are much lower than vacuum-tube circuits (6 to 8 V compared to 90 to 100 V). If vacuum-tube-type horizontal checking equipment is used to test solid-state circuits, the components will probably be damaged. In any event the results will prove nothing. There are horizontal test sets manufactured specifically for solid-state TV. If these test sets are not available, you must rely on waveform and voltage measurements.

The most common causes of trouble in the horizontal output and high-voltage circuits are (in order) capacitors, horizontal output transistor (since it operates at high current), diodes, and transformers.

Since solid-state horizontal circuits do not operate in the same way as do the corresponding vacuum-tube circuits, the same symptoms do not always mean the same trouble. For example, the horizontal output transistor is in

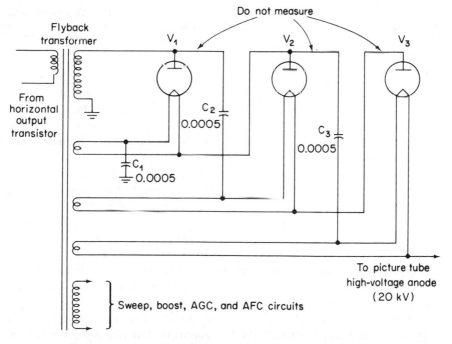

Figure 6–6. Voltage doubling circuit found in high-voltage section of some hybrid sets.

effect a diode in parallel with the damper diode. If the damper diode is open, the transistor can provide the same function and act as a damper. Of course, picture linearity will be very poor, especially on the *left side* of the screen. A shorted damper diode will usually make the entire circuit inoperative.

In many vacuum-tube circuits the picture-tube boost voltages are supplied by the damper. If the boost voltage is present, the damper is good. In most solid-state circuits the boost is supplied by a separate diode, so the presence or absence of boost voltage has no bearing on the condition of the damper diode. Also in most solid-state circuits, it is possible for a normal high voltage to be present, even though there is an open in the horizontal deflection yoke. Of course, there will be no horizontal sweep (only a vertical trace on the picture-tube screen) or the horizontal sweep will be impaired.

In some hybrid sets a voltage doubler circuit is used, as shown in Fig. 6–6. This is found mostly in large-screen (23–in.) sets where the high voltage must be about 25 to 30 kV. Note that V_1 and V_3 conduct on positive peaks, while V_2 conducts on negative peaks. Capacitors C_1, C_2, and C_3 remain charged on both positive and negative peaks. This produces a d-c output more than double that available across the transformer winding.

6–2.2. Typical Troubles

The following paragraphs discuss symptoms that can be caused by defects in the high-voltage supply and horizontal output circuits.

6–2.2.1. Dark Screen

If the picture screen is dark but sound is normal, it is likely that the horizontal output and high-voltage circuits are at fault. Of course, the picture tube could be defective or the problem could be something simple such as a lack of filament voltage. Picture-tube troubles are discussed in a later section.

The most likely defects are open, shorted, or leaking capacitors; leaking horizontal output transistor; open or shorted diodes; and defective flyback transformer.

Before checking individual parts, there are some circuit tests that will help isolate the problem. These are as follows: check drive voltage waveform at the horizontal output transistor base and the sweep output at the collector; check for high voltage at the picture-tube second anode, observing precautions discussed previously; check for boost voltage and any auxiliary voltages (focus, accelerator grid, AFC or AGC pulses, etc.).

The first check to be made depends on whatever is the most convenient. For example, it is logical to check the horizontal output transistor waveforms before checking high voltage. In some sets, the transistor may be at a very inaccessible location, so the best bet is to check the high voltage first.

If the drive waveform at the horizontal transistor base is absent or abnormal, the problem is ahead of the horizontal output circuits. If the base waveform is good but the collector waveform is absent or abnormal, the transistor is the first suspect. If the collector waveform is good but one or more of the output voltages is bad, check individual parts in the related circuit. Also check the d-c voltages at each of the transistor elements.

Shorted capacitors (a common problem in the case of a dark screen) will show up when d-c voltages are measured. Open or leaking capacitors are not located as easily. Generally, an open or leaking capacitor will produce an abnormal waveform. If a capacitor is suspected, try lifting one lead and checking the capacitor with a tester (for leakage, value, etc.), or try substitution if it is convenient. Again, it is not recommended that capacitors be checked by shunting with a good capacitor, unless the power is first turned off. The voltage surges can be damaging to transistors.

Leaking transistors will show up when waveforms are measured. The problem is confirmed further when the d-c voltages are measured. Collector-base leakage in the horizontal output transistor is a common problem. If leakage is bad enough to cause complete failure of the circuit which results in a dark screen, the d-c voltages at the transistor elements will be incorrect. Keep in mind that the horizontal output transistor operates at or near zero bias, or possibly with reverse bias. Therefore, if there is any substantial forward bias, it is probably the result of leakage. Any collector-base leakage will forward-bias a transistor. In the case of a normally cutoff horizontal output transistor, the undesired forward bias will attenuate the collector waveform in addition to producing incorrect d-c voltages.

If the capacitors and transistor appear to be in order, check the diodes. If the sweep output waveform (transistor collector) is abnormal, check the damper diode. If the output voltages are abnormal (with a good sweep output), check the corresponding rectifier diodes.

A shorted damper diode is usually easy to pinpoint, since the transistor collector waveform and d-c voltage will be abnormal. An open damper diode will usually not provide a dark screen (total failure of the circuit). If any of the other diodes are defective, this will show up as an absent or abnormal output voltage.

Note that in some all-solid-state circuits, such as shown in Fig. 6–4, the high-voltage rectifier is actually a group of several series-connected diodes. If any one of these diodes becomes shorted or develops excessive leakage, the remaining diodes can break down. This is because the normal voltage drop across the defective diode is placed on the remaining diodes, resulting in abnormally high peak voltage across the other diodes.

If the flyback transformer has an open winding or if a large portion of a winding is shorted, the problem is usually self-evident. However, if there is only a partial short, leakage between windings, or a high-voltage arc, it may be difficult to check. Substitution is the only sure check, unless you have a flyback transformer tester suitable for solid-state circuits. Unfortunately, replacement of the flyback transformer is not an easy job. Therefore, do not try substitution except as a last resort (when capacitors, diodes, and transistor all prove to be good).

6–2.2.2. Picture Overscan

Picture overscan or "blooming" in solid-state sets is the same as in vacuum-tube TV. The picture becomes dim even with the brightness control full on, and there is an abnormal enlargement of the picture or raster. Usually the enlargement is uniform, but there may be some defocusing. In some circuits, the brightness control operates in reverse—rotating the control for an increase in brightness produces a decrease—after reaching a critical point on the control. If picture overscan is not accompanied by insufficient width (a symptom of horizontal output failure), the probable cause is a failure in the high-voltage circuit only.

Any circuit problem that reduces (but does not completely eliminate) the high-voltage output to the picture-tube second anode can cause overscan. The most likely defects are leaking high-voltage capacitors, defective high-voltage tube or rectifiers, leaking high-voltage leads, defects in high-voltage circuit protective resistances (if any), and shorted turns in the high-voltage winding of the flyback transformer.

The first obvious test is to measure the high voltage (observing all precautions). If the high voltage is normal (unlikely), try a new picture tube. Also, there may be a corona problem (leakage from the high-voltage lead) especially in large-screen TV, just as there is in vacuum-tube sets. The only practical cure is to replace the lead.

If the high voltage is low, check any high-voltage filter capacitors for leakage. In a hybrid set, try replacing the high-voltage tube. If the high-voltage circuit has protective resistors (R_3 and R_5, Fig. 6–3), as is the case in many hybrid circuits, check the resistance values. Resistor R_3 protects the high-voltage system in case of a short. Resistor R_5 sets the high-voltage rectifier tube filament to the correct value. If either resistor increases in value (say, due to overheating), the high-voltage output will be reduced.

It should be noted that many *color TV* pictures and rasters bloom when *both* brightness and contrast controls are turned full up. This is an inherent design problem (lack of high-voltage regulation).

6–2.2.3. Narrow Picture

The most logical cause of a narrow picture that cannot be corrected by adjustment of the width controls is insufficient horizontal drive. This is usually accompanied by other symptoms, such as decreased brightness, picture distortion, and the like. Very often, a narrow picture (insufficient drive) is the result of marginal breakdown rather than a complete breakdown. For example, if the horizontal output transistor has some collector-base leakage, the transistor will be forward-biased and the sweep output will be decreased.

The most likely defects depend on symptoms that accompany the basic problem of a narrow picture. For example, if there is distortion on the left-hand side of the picture, look for an open or leaking damper diode. If there is right-hand distortion, check for a defective horizontal output transistor.

The first test is to measure the horizontal output transistor collector waveform, as well as the sweep waveform to the horizontal yoke. These waveforms will rarely, if ever, be normal when there is insufficient picture width.

Note that in the circuit of Fig. 6-4, the horizontal transistor collector is connected directly to the horizontal yoke. In other circuits, such as shown Fig. 6-3, the horizontal yoke is supplied by a winding on the flyback transformer.

The waveform measurements should be followed by voltage measurements at the horizontal output transistor elements. Abnormal voltage readings will show up such defects as leaking or shorted capacitors and diodes. As a final resort, look for a marginal breakdown in the flyback transformer, such as a partially shorted winding or some leakage between windings.

Do not overlook insufficient horizontal drive, especially when waveforms and voltages all appear normal but somewhat low. Check the drive waveform at the horizontal output transistor base. Look particularly for a drive waveform (at the base) that is normal but low in amplitude (below the usual 6 to 8 V). Compare the base and collector waveform amplitude and the horizontal yoke waveform amplitude very carefully against those shown in the service literature. In solid-state circuits, even a slight drop in waveform amplitude can indicate a large change in current (which can produce major problems in current-operated solid-state circuits).

6-2.2.4. Foldback or Foldover

Horizontal foldback usually occurs only on one side of the picture screen. A portion of the picture is folded back on one edge of the display. In some rare cases, there will be a fold in the center.

Assuming that the horizontal drive signal is normal, the problem can be traced to the horizontal sweep components. The most likely defects are horizontal output transistor, damper diode, damper capacitors, horizontal yoke, or yoke drive winding on the flyback transformer. If foldback is on the right-hand side, look for a defective transistor or its related components. If foldback is on the left-hand side, check the damper diode and its related components.

The first test is to measure both the base (drive) and collector (output) of the horizontal output transistor. Then measure the horizontal yoke waveform (if it is different from the collector waveform). Invariably, the yoke waveform will be distorted.

Keep in mind that the horizontal sweep system is essentially a resonant circuit. The inductance of the flyback transformer combines with various capacitors in the circuit (such as C_1 in Fig. 6–3 and C_4 in Fig. 6–4) to resonate at about 50 kHz (typical). If the resonant frequency is not correct (generally low), the waveform will be distorted, resulting in foldback.

If foldback occurs after replacement of any part in the horizontal system, the problem is almost certainly one of an incorrect value: out-of-tolerance capacitor; flyback transformer with incorrect winding inductance (d-c resistance may be good); or a yoke with incorrect inductance. If foldback occurs before replacement of parts, look for parts that have gone out of tolerance. In solid-state horizontal sweep circuits, the maximum tolerance is usually 10 per cent, with 5 per cent producing better results.

In extreme cases, when all parts in the system appear to be good but foldback continues, check the duration of the drive pulse (*on* time compared to *off* time) against the service literature.

6–2.2.5. Nonlinear Horizontal Display

Any nonlinearity in the horizontal display is almost always found with at least one other problem [narrow picture, overscan (blooming), lack of brightness, foldback, etc.]. Therefore, the most likely causes of nonlinearity are the same as for the other symptoms. Also, the troubleshooting test sequence should be the same (waveform measurements followed by voltage measurements).

Horizontal nonlinearity should not be confused with the *keystone* effect. Horizontal keystoning is evident when the top and bottom widths of the picture display are not the same (picture is wider at the top than at the bottom or vice versa) and is almost always caused by a problem in the horizontal yoke. (One set of horizontal deflection coils has shorted turns and is unbalanced with the other set of coils.)

6–3. HORIZONTAL OSCILLATOR AND DRIVER

The horizontal oscillator and driver circuits provide the drive signal to the horizontal output and high-voltage supply section. These signals are at a frequency of 15,750 Hz and are synchronized with the picture transmission by means of sync signals (taken from the sync separator section, as described in later paragraphs). Most solid-state horizontal oscillator

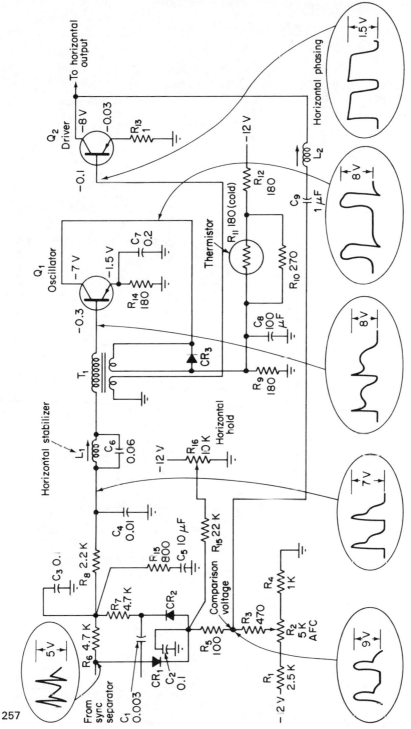

Figure 6–7. Horizontal oscillator and driver with balanced AFC.

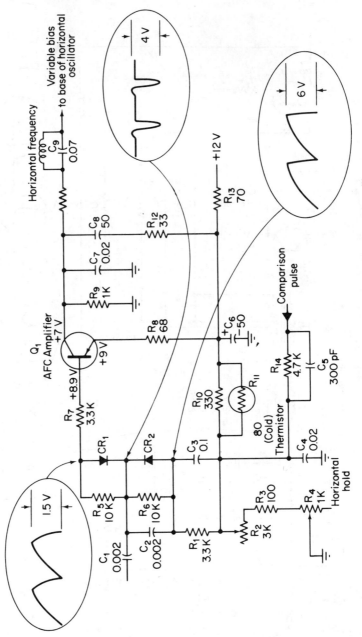

Figure 6–8. Unbalanced AFC circuit for horizontal oscillator.

circuits include some form of AFC system to ensure that the horizontal sweep signals are synchronized (both for frequency and phase) with picture transmission, despite changes in line voltage and temperature or minor variations in circuit values. The AFC action is accomplished by comparison of the sync signals with horizontal sweep signals, both for frequency and phase. Deviations of the horizontal sweep signals from the sync signals cause the horizontal oscillator to shift in frequency or phase as necessary, to offset the initial (undesired) deviation. For example, if the horizontal sweep increases in phase from the sync signals, the horizontal oscillator is shifted in phase by a corresponding amount but in the opposite direction (a decrease in phase) to offset the change.

There are three basic horizontal oscillator circuits used in solid-state TV. These include the balanced AFC (Fig. 6–7), the unbalanced AFC (Fig. 6–8), and the transistor AFC (Fig. 6–9). In all cases the circuit is composed of three sections: the horizontal AFC, the horizontal oscillator, and the hori-

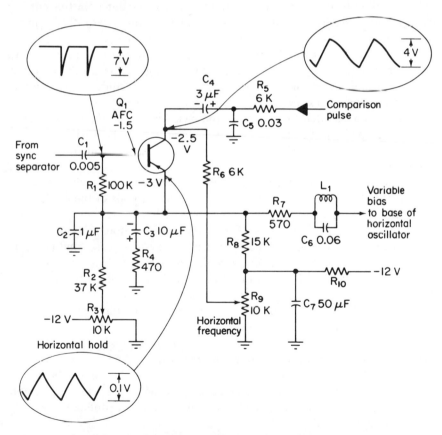

Figure 6–9. Transistor-type AFC circuit for horizontal oscillator.

zontal driver. In a few cases the horizontal circuits are more elaborate. For example, there may be a buffer or buffer/amplifier transistor stage between the horizontal oscillator and the driver. Also, there may be a phase inverter stage between the sync separator and the horizontal AFC. Generally, however, the buffer stage is omitted, and the horizontal phase inversion (if any) is part of the sync separator section.

In all cases the horizontal oscillator is of the *blocking oscillator* type, operating at a frequency of 15,750 Hz and producing an output of about 1 to 3 V. The output is amplified to about 6 to 8V by the horizontal driver (a basic, common-emitter amplifier) and is applied to the horizontal output section. In some cases the horizontal driver output is as high as 50 V.

As in the case of most blocking oscillators, the horizontal oscillator frequency is determined by circuit values and by the bias voltage. In no case is the horizontal oscillator triggered directly by the sync pulses. Instead, the sync pulses and comparison pulses produce a variable d-c control voltage that is applied to the horizontal oscillator base. Any change in this control voltage shifts the horizontal oscillator frequency and phase. The control voltage is manually set by the horizontal adjustment controls (horizontal frequency, AFC, horizontal hold, etc., depending on the type of circuit) to an average value. Any deviation from the average value produces a corresponding change in the base bias which, in turn, shifts the horizontal oscillator frequency as necessary.

In the balanced circuit of Fig. 6–7, the base bias for horizontal oscillator Q_1 is developed by the AFC diodes CR_1 and CR_2. The variable control voltage is filtered by an *RC* circuit with a long time constant. Any random noise pulses that may be mixed with the horizontal sync pulses are averaged out by the *RC* network, thus rejecting noise interference as much as possible.

Note that the sync pulses are applied to opposite ends of the diodes CR_1 and CR_2, while the comparison pulses and the horizontal control voltages are applied to the diode common (CR_1 cathode, CR_2 anode). Diode conduction depends on the combined peak voltages of the sync pulses and the comparison pulses. The balanced AFC circuit requires a double-ended or push/pull input from the sync separator and is similar to a ratio detector used in FM detectors.

In the unbalanced circuit of Fig. 6–8, the input from the sync separator is single ended, while the control voltage output is applied through an amplifier Q_1. The AFC diodes CR_1 and CR_2 receive both sync pulses and comparison pulses. Diode current depends on the instantaneous voltage of the two waveforms and thus changes from one instant to another. However, there is an average value of current in each diode over one complete cycle of operation. The amount of d-c voltage produced is determined by the difference between these two average values.

The *relative phase* of the two waveforms across each diode changes if the horizontal oscillator drifts off frequency. Thus, the average value of rectified voltage also changes. One diode conducts more than the opposite diode, and the d-c control voltage varies higher or lower, depending on the phase error (horizontal oscillator leading or lagging the sync pulses).

In the circuit of Fig. 6–9, transistor Q_1 is used in place of the AFC diodes. The Q_1 base is driven by the sync pulses, while the collector receives comparison pulses. The emitter waveform is a combination of both pulses, and when there is a change in phase between the two sets of pulses, the d-c emitter current changes. In turn, the d-c control voltage applied to the horizontal oscillator changes in a direction that corrects the frequency and phase of oscillation.

6–3.1. Recommended Troubleshooting Approach

Failure of the horizontal oscillator and related circuits (AFC and horizontal drive) can produce symptoms similar to those produced by failure in other circuits. For example, if the horizontal oscillator stops oscillating, there will be no drive signal to the horizontal output circuit. Thus, there will be no high voltage or boost voltage, and the picture-tube screen will be dark. This same symptom can also be produced by failure of the horizontal output circuit, the low-voltage power supply, and the picture tube.

Also, operation of the horizontal oscillator circuits depends on signals from other circuits. For example, the horizontal AFC circuit must have sync signals from the sync separator and comparison signals from the horizontal output to operate properly.

Because of these conditions, the only practical troubleshooting approach for the horizontal oscillator circuits is to measure waveforms at all outputs and inputs, followed by d-c voltage measurements at all transistor elements.

If the driver output is absent or abnormal, with good sync and comparison pulse inputs, the problem is definitely in the horizontal oscillator or driver sections. A waveform measurement at the input of the horizontal driver (transistor base) or horizontal oscillator output will localize the problem further.

If the driver output is present but the symptoms point to a horizontal circuit failure (horizontal pulling, jitter, distortion, etc.), the problem is most likely in the AFC circuits. Problems in the AFC section are usually isolated by means of voltage measurements followed by substitution of parts.

Note that the horizontal oscillator often appears to be reverse-biased, on the basis of d-c voltage measurements. However, it is still possible for the oscillator to be forward-biased during the heavy collector current pulses (*on* time) and then drops back to reverse bias between pulses (*off* time). Therefore, the only test of the horizontal oscillator is the presence of correct wave-forms at both the base and collector or the output winding of the horizontal oscillator transformer.

Also note that the horizontal stabilizer and horizontal phasing controls (if any) will have little effect on the d-c voltages but will have considerable effect on waveforms. On the other hand, the potentiometer controls such as the horizontal frequency and horizontal hold will have considerable effect on the d-c voltages. For example, in Fig. 6–7, the horizontal hold potentiometer will vary both the base and emitter voltages of the horizontal oscillator by more than 0.5 V. The presence or absence of sync pulses will mean less than about 0.3-V difference in d-c voltages at the base and emitter of the horizontal oscillator.

The horizontal drive is usually zero-biased or slightly forward-biased. The horizontal oscillator output then drives the driver into heavy conduction, possibly to saturation. In circuits that use a transistor for the AFC (see Figs. 6–8 and 6–9), the transistor is reverse-biased.

6–3.2. Typical Troubles

The following paragraphs discuss symptoms that could be caused by defects in the horizontal oscillator circuits.

6–3.2.1. Dark Screen

If the picture screen is dark (no raster) but sound is normal, indicating that the low-voltage power supply is good, the first step is to measure the waveform at the horizontal drive output. If the waveform is normal, the problem is in the horizontal output and high-voltage circuit or in the picture tube itself. If the waveform is not normal (weak, distorted, etc.), the problem is probably in the horizontal oscillator. The most likely defects are open, shorted, or leaking capacitors; leaking transistors; open or shorted diodes; and defective blocking oscillator transformer.

Before checking individual parts, there are some circuit checks that will help isolate the problem. These are waveforms at sync input (from sync separator); comparison pulse input; horizontal oscillator base and collector (and/or output winding of blocking oscillator transformer); and voltages at all transistor elements.

If the sync pulses are absent or abnormal, the problem is in the sync separator rather than in the horizontal oscillator section. If the comparison pulses are not normal, with a good output from the horizontal drive, then the problem is in the horizontal output and high-voltage section.

If the sync pulses and comparison pulses are both normal but the horizontal oscillator output is absent or abnormal, the problem could be in the

AFC section or the horizontal oscillator. If the horizontal oscillator output is normal, the problem is in the horizontal drive circuit.

As usual, capacitors are the most likely cause of trouble in the horizontal oscillator circuits. For example, if C_8 in Fig. 6–7 is shorted, Q_1 will have no collector voltage and will stop oscillating. The same is true if there is a short in capacitor C_4. In this case, the d-c control voltage from the AFC section will be shorted to ground, instead of biasing the horizontal oscillator base to the correct level.

If the TV set is old, look for worn potentiometer controls. For example, if R_2 in Fig. 6–7 is open, the bias on Q_1 will be abnormal. Q_1 may continue oscillating, but the frequency and amplitude will be off, probably both low. This will show up as an abnormal waveform and as incorrect voltages on the transistor elements.

If the capacitors and controls appear to be good, look for a defective diode. For example, if either CR_1 or CR_2 is defective, the control voltage on the base of Q_1 will be abnormal. Likewise, a shorted CR_3 will prevent the collector-to-base feedback necessary for Q_1 oscillation. Diodes can be checked with an ohmmeter, as described in Chapter 2. However, always lift one lead before making the check. If CR_3 is measured in circuit, it will appear shorted since the ohmmeter will measure the transformer winding resistance (typically less than 5 Ω).

6–3.2.2. Narrow Picture

The most logical cause of a narrow picture that cannot be corrected by adjustment of the width controls is insufficient horizontal drive. If the problem is in the horizontal oscillator circuits, it will show up as a low output waveform from the horizontal driver and/or horizontal oscillator.

The most likely defects are open emitter capacitor in the horizontal oscillator, collector-base leakage, shorted turns in the blocking oscillator transformer, and possibly off-value resistors. However, any number of defects could produce a low output.

The first step is to isolate the problem to the horizontal driver or the horizontal oscillator by waveform measurement. If the problem is in the horizontal oscillator, keep in mind that the trouble can be in the AFC section.

6–3.2.3. Horizontal Pulling or Improper Phasing; Loss of Sync

Horizontal pulling is present when the picture pulls and appears in diagonal form. If the picture pulls completely into diagonal lines, this indicates a complete loss of sync. Note that the direction of the slant can provide a clue to the problem. If the lines slant to the right, the horizontal oscillator frequency is high. If the lines slant to the left, the frequency is low.

If the picture shifts to the right or left so that it is decentered, this indicates incorrect phasing. The horizontal oscillator is on frequency but not in phase with the sync signals.

It is possible that any of these problems can be the result of improper adjustment of controls. Therefore, the first step is to adjust all of the horizontal controls. If this does not clear up the problem, check all of the waveforms and transistor voltages. Pay particular attention to the sync pulses (from the separator) and the comparison pulses (from the horizontal output circuits). If either of these are absent or abnormal, the AFC circuits will not operate properly.

For example, if the comparison pulses are absent, horizontal sync may not be completely lost. However, the horizontal hold adjustment will become very critical, and phase shift will occur (picture will be decentered to the left).

If the sync and comparison pulses are normal, the most likely defects are capacitors and/or diodes in the AFC section. However, there are many other defects that can cause these symptoms, so an analysis of the symptoms and measurements (waveform/voltage) must be made before checking individual parts.

Defects in the AFC diodes (CR_1 and CR_2 in Fig. 6–7) are quite common. Complete shorts or opens in these diodes will show up as severe distortion of waveforms or transistor voltages that are well out of tolerance. However, marginal defects in the AFC diodes can be difficult to locate. For example, a poor front-to-back ratio can result in a touchy horizontal hold.

If the picture pulls at the top, try adjustment of the horizontal stabilizer control (L_1 in Fig. 6–7). If the adjustment has no effect on pulling, this points to a defect in the AFC circuits. Also try adjustment of the horizontal stabilizer control in the event of horizontal sync loss. If the horizontal hold control is very touchy and the images appear to overlap, it is possible that the stabilizing capacitor (C_6 in Fig. 6–7) is open. This is confirmed if the stabilizer control has little effect on the waveform at the base of the horizontal oscillator.

6–3.2.4. Horizontal Distortion

There are many forms of horizontal distortion that can be caused by defects in the horizontal oscillator and driver circuit. The so-called "pie-crust" distortion is a typical example. With pie-crust distortion, the picture image appears to be made up of wavy lines, even though there is no pulling, loss of sync, or jitter. In solid-state TV, any form of distortion is almost always the result of marginal performance in a particular component rather than complete failure.

The most likely defects are in capacitors, particularly the filter capacitors that filter the variable d-c control voltage from the AFC diodes to the horizontal oscillator. For example, a slight leakage in capacitors C_3, C_4, and C_5 in Fig. 6–7 can cause horizontal distortion. Transistor leakage is the next logical suspect. However, such leakage will usually cause other symptoms to appear (pulling, loss of sync, etc.).

6–4. VERTICAL SWEEP CIRCUITS

The vertical sweep circuits provide a vertical deflection voltage (vertical sweep) to the picture-tube deflection yoke. The vertical circuits also supply a blanking pulse to the picture tube (usually through the video amplifier, as discussed in later paragraphs). This pulse blanks the picture tube during retrace of the sweep.

The vertical sweep signals are at a frequency of 60 Hz and are synchronized with the picture transmission by means of sync signals (taken from the sync separator section, as described in later paragraphs).

Figure 6–10 is the schematic diagram of a typical vertical sweep circuit. Note that there are three stages (oscillator, driver, and output). In some solid-state TV sets, the driver and output functions are combined into one stage.

The vertical oscillator is of the blocking oscillator type, operating at a frequency of 60 Hz and producing a pulse output of about 1 to 2 V. The oscillator pulse output is modified into a sawtooth sweep and is applied through the driver to the output stage. The final circuit output is a peaked sawtooth waveform. The sawtooth portion (about 4 to 5 V) is applied to the picture-tube deflection yoke, while the peaked portion (about 50 V) is used as the blanking pulse for the picture tube.

The vertical oscillator frequency is determined by circuit values and by the bias voltage. However, the oscillator is locked in frequency by the sync pulses. As shown in Fig. 6–10, Q_1 is normally reverse-biased where d-c voltages are concerned. Feedback is obtained through coupling of the collector and base windings of T_1. Whenever Q_1 conducts, the winding feedback makes the base more negative, increasing the collector conduction and charging C_1 to a high negative value. When the emitter of Q_1 goes sufficiently negative, Q_1 is cut off and remains so until C_1 discharges through R_1.

As with most solid-state blocking oscillators, Q_1 is on (conducting) for a very short period of time and is off (nonconducting) for a long period of time during each cycle. The point at which Q_1 starts to conduct (and thus the frequency of Q_1) is set by vertical hold control R_2 which controls the base bias. Actually, R_2 is set so that the frequency of Q_1 is just below 60 Hz. In practice, R_2 is adjusted until the vertical sweep is locked with the picture transmission. That is, the sync pulses trigger Q_1 into conduction *earlier* than Q_1 would otherwise start to conduct.

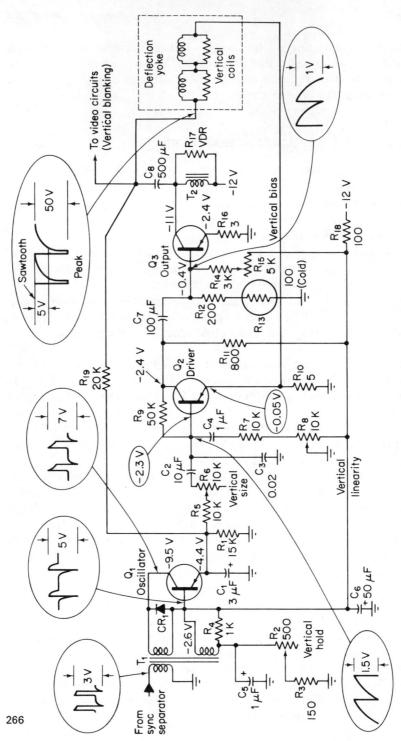

Figure 6–10. Vertical sweep circuits.

Diode CR_1 is often a source of trouble in the circuit. CR_1 acts as a blocking diode to prevent the Q_1 signal from going back into the sync separator. CR_1 also prevents an excessive peak voltage on the base of Q_1. If CR_1 opens, Q_1 can be destroyed. If CR_1 is shorted, Q_1 will not oscillate. If CR_1 is otherwise defective, the Q_1 pulses can feed back into the sync separator and disturb its operation.

In the circuit of Fig. 6–10, the Q_1 pulse output is converted to a sawtooth sweep by means of the emitter network (C_1R_1). The discharge of C_1 through R_1 produces the basic sweep waveform. The sweep is made linear through feedback of the output by R_{19}. The waveform is also shaped by the network $C_3C_4R_7R_8$. Vertical linearity control R_8 provides a manual control for the sawtooth waveform.

The amplitude of the sawtooth waveform to driver Q_2 (and thus the vertical size of the picture raster) is set by vertical size control R_6. Note that the emitter of Q_2 is connected to the deflection yoke. This provides negative feedback to ensure a more linear output.

The deflection yoke of a solid-state TV requires heavy current, even though the sweep voltage is low. For this reason, the output transistor Q_3 is of the power type (often mounted on a heat sink). Since there is always the danger of thermal runaway in power transistors, thermistor R_{13} is included in the base circuit of Q_3. If collector current increases, heating Q_3, and causing more current flow, the resistance of R_{13} will drop, lowering the emitter-base voltage differential. In turn, this will lower the emitter-base and emitter-collector current flows. Transistor Q_3 is further stabilized by unbypassed emitter-resistor R_{16}. If it becomes necessary to replace transistor Q_3, it may be necessary to adjust vertical bias control R_{15}.

Most solid-state vertical output circuits have a bias control, since the characteristics of replacement transistors can be different from the original transistor. The peaked portion of the output waveform is obtained by inductive kickback of T_2. Typically, the peak will be about 50 V, even though the sawtooth portion is usually about 5 V.

Resistor R_{17} is a voltage-dependent resistor (VDR) found on many late-model sets. Resistor R_{17} keeps the amplitude of the output waveform within narrow limits. The resistance of a VDR is varied by the output voltage to the deflection yoke. If the output tends to increase, the resistance of R_{17} decreases, keeping the output constant. Resistor R_{17} should be checked first if the trouble symptom is a constantly varying vertical size or vertical height.

6–4.1. Recommended Troubleshooting Approach

Complete failure of the vertical sweep circuits is an easy symptom to recognize and is usually easy to locate. With complete failure, there will be no vertical sweep, as the picture display will be a horizontal line only. (As in the case of vacuum-tube sets, do not operate solid-state TV with a failure in the vertical sweep. The bright horizontal line can burn out the picture-tube screen.)

In the case of complete vertical sweep failure (all other functions normal), the logical first step is to check waveforms at the vertical oscillator output (driver input), driver output, and power output. If the vertical oscillator waveform is absent or abnormal, the problem is quickly localized to the vertical oscillator stage. (Note that in some solid-state vertical circuits the oscillator will not go into oscillation unless there are sync pulses present. Therefore, it is always wise to check for sync pulses if the oscillator appears defective.) If the oscillator waveform is normal but either the driver or power output waveforms are abnormal, this will localize the problem to these stages. If all of the waveforms appear normal but there is no vertical sweep, the deflection yoke is a logical suspect.

Marginal failure of the vertical sweep circuits is not so easy to recognize or locate. Such problems as loss of vertical sync (or very critical vertical sync), distortion (nonlinearity), line pairing or splitting (poor interlace), or lack of vertical height can be caused by any number of defects in the vertical sweep circuits.

In the case of marginal failure, the first logical step is to try correcting the problem by adjustment of the vertical controls. If the problem cannot be eliminated by adjustment of controls, or if it is necessary to set a control to an extreme (one end or the other), then a marginal component failure can be suspected. Leaking capacitors, transistors, and diodes, or worn potentiometers are likely causes. An open capacitor can also be a cause.

The next logical step in marginal failure is comparison of waveforms and/or voltages against those shown in the service literature. Usually, waveform measurement will be of the greatest value. One point to consider in making waveform measurements in solid-state vertical sweep circuits is possible false distortion. The sawtooth waveform developed by the vertical sweep circuits must be linear to get a linear vertical picture. A poor oscilloscope (one with narrow bandwidth or excessive input capacitance) can indicate a nonlinear sweep when a good sweep is present. Also, the oscilloscope can load the circuit and create nonlinearity.

6–4.2. *Typical Troubles*

The following paragraphs discuss symptoms that could be caused by defects in the vertical sweep circuits.

6–4.2.1. *No Vertical Sweep*

If the picture display consists of only a bright horizontal line, indicating that all circuits but the vertical sweep are probably normal, the first step is to measure waveforms in the vertical sweep circuits. Measure all of the waveforms (or their equivalents) shown in Fig. 6–10. Then measure the voltages at all of the transistor elements.

It is very unlikely that there will be a complete loss of vertical sweep without at least one of the waveforms and/or voltages being abnormal. A possible exception to this is where the vertical coils of the deflection yoke are open. Always check the yoke windings if the final output (peaked sawtooth) waveform appears normal but there is no vertical sweep.

Always compare the transistor voltages against the service literature, when available. If literature is not available, use the following rule of thumb: the vertical oscillator and vertical output transistors will be reverse-biased, while the driver is usually forward-biased. In two-stage vertical sweep circuits, the output transistor is zero-biased or slightly forward-biased, with the oscillator being reverse-biased.

Keep in mind that the reverse-bias condition of the vertical oscillator Q_1 is caused by the charge built up across the emitter capacitor C_1. If it were not for this charge build-up, Q_1 might be forward-biased (as it is during the brief *on* time). However, with Q_1 operating normally (oscillating), the average base-emitter differential will be such that Q_1 appears to be reverse-biased. Any solid-state blocking oscillator that shows a forward-biased condition is suspect.

If it becomes necessary to replace the vertical output transistor, be sure to follow the precautions for power transistors outlined in Sec. 2–10.1.2. Also, if the power transistor is provided with a protective thermistor, check the thermistor whenever the power transistor is replaced. Generally, if the cold resistance of a termistor is within tolerance, the thermistor is good. Sometimes, a thermistor can show a good cold resistance but still be defective when heated. As a precaution, check the base-emitter bias of the power transistor when power is first applied and as the circuit warms up. If the base-emitter bias voltage does not stabilize and the power transistor appears to overheat, remove the power and replace the thermistor.

6-4.2.2. Insufficient Height

The lack of vertical height can be caused by improper adjustment of controls or by defective circuits. Therefore, the first step in troubleshooting this symptom is to adjust the vertical height control (sometimes called the vertical drive control or the vertical size control). There is no problem if proper height can be obtained with the vertical height control at midrange (or not more than about 3/4-range). If the vertical height control must be on full, or near full, to get the proper height, it is possible that the vertical output transistor is improperly biased. If the vertical output transistor is biased near the cutoff point, part of the sawtooth sweep will be clipped off, reducing the output. Therefore, the next step in troubleshooting is to adjust the vertical output transistor bias.

Follow the service literature instructions, if available. If not, adjust the bias control to produce the correct waveform amplitude at the vertical output transistor collector. In the circuit of Fig. 6–10, the sawtooth sweep is about 5 V, while the pulse peaks are about 50 V.

If there is no service literature, try the following procedure to set the vertical output bias. Set the vertical height control to mid range. Then adjust the bias control until the picture is at the desired height. If you get the correct height without distortion or other symptoms of vertical circuit failure, both the height and bias controls are properly set.

If it is impossible to get the proper vertical height by adjustment of the controls, the most likely causes are leaking capactiors, worn vertical height control, and leaking transistor (particularly the output transistor).

Measure all waveforms and d-c voltages in the vertical sweep circuits. Look particularly for low amplitude in the sawtooth sweep at any point in the circuit. Aside from adjustments, the only logical causes for lack of vertical height are low amplitude sweeps or a defective yoke. If the final output sweep is of the correct amplitude, check the yoke. If the final output sweep is low, work back to the vertical oscillator.

It is fairly certain that one or more of the waveforms and/or voltage measurements will be abnormal if there is insufficient vertical height. However, the abnormal indication may not pinpoint the defective component, particularly in the case of a marginal defect. The following are some typical examples using the circuit of Fig. 6–10.

If the vertical oscillator waveform is low, look for leakage in the bypass capacitor C_1 or the protective diode CR_1. If the drive waveform is low, look for leakage in wave-shaping capacitor C_3. If the output waveform is low (with the yoke good), look for a defective VDR R_{17} (if used) or a defective output transformer T_2. Usually, it will be necessary to check T_2 by substitution, since a marginal defect such as shorted turns or leakage between windings may not show up in resistance measurements.

Do not overlook the possibility of leakage in the output transistor Q_3. Any collector-base leakage will reduce the output. Keep in mind that it is possible for a transistor to be leaking and still show the correct base-emitter differential voltage. However, both voltages will be abnormally high in relation to ground or common. This is because collector-base leakage tends to forward-bias the transistor, resulting in more emitter current flow and a larger drop across the emitter resistor.

6-4.2.3. Vertical Sync Problems

The lack of vertical sync or a very critical vertical sync can be caused by improper adjustment of controls, as well as by defective circuits. Therefore, the first step in troubleshooting this symptom is to adjust the vertical hold (sometimes called vertical lock or vertical sync) control.

There is no problem if vertical sync can be obtained with the vertical hold control at midrange. If the vertical hold control must be full on (or near full) or it is necessary to readjust the control frequently, there is a problem in the vertical sync circuits.

If the problem cannot be cured by adjustment, the next step is to localize the fault. The same symptoms can be caused by a defect in the vertical portion of the sync separator (as discussed in later paragraphs).

First, check for the presence of proper sync pulses at the input to the vertical sweep circuits (at the primary of T_1 in Fig. 6–10). Then, check the sawtooth sweep portion of the vertical oscillator (say, at the base of Q_1). With the picture rolling, the vertical sync pulse will appear to ride on the sawtooth, as shown in Fig. 6–11.A. Normally (picture not rolling), the sync pulse is hidden since it is close to the beginning (or end) of the sawtooth sweep.

If the input sync pulses are not normal, the problem is in the sync separator. If the sync pulses do not appear in the vertical oscillator sawtooth waveform (with picture rolling), the problem is likely to be in the input transformer T_1 or related circuits.

With the picture still rolling, check the rate of roll. If the roll is steady, the vertical oscillator is probably good but the sync pulses are not of sufficient amplitude to lock the oscillator in sync. (Or the oscillator is biased such that the sync pulses cannot overcome the reverse bias.) If the roll rate is unsteady with constant sync pulses of sufficient amplitude, the problem is in the vertical oscillator.

If the sync pulses are normal and it is impossible to get proper vertical sync, the most likely causes are leaking bypass capacitors (C_5 and C_6 in Fig. 6–10), worn vertical hold control, leaking or defective vertical oscillator transistor, or defective vertical oscillator transformer (leakage between windings or some similar marginal defect).

If bypass capacitors are leaking, the vertical hold control will be partially shorted; it will not have its full resistance insofar as the circuit is concerned. This same condition will result if the vertical hold control has a worn resistance element. In the case of a very critical vertical sync (difficult to adjust or will not hold after adjustment), observe the amplitude of the vertical sync pulse riding on the sawtooth portion of the vertical oscillator waveform. If the sync amplitude does not remain constant, look for problems in the sync separator rather than in the vertical sweep circuit.

6-4.2.4. *Vertical Distortion (Nonlinearity)*

There are several forms of vertical distortion. Some are easy to recognize, such as *keystoning* where one side of the picture is much larger than the opposite side. Keystoning is generally caused by a defect in the deflection yoke or related circuit parts (such as a thermistor in the yoke circuit) but could be caused by a marginal defect in the vertical output transformer (T_2 in Fig. 6–10).

Other forms of vertical distortion are not as easy to recognize. In the extreme, there will be a compression of the picture at the top with picture spreading at the bottom, or vice versa. Often, the compression or spreading is slight. Use a test pattern transmission or a cross-hatch pattern from a TV signal generator to check on linearity. If neither of these are available, a quick check of vertical linearity can be made as follows. Adjust the vertical hold control to produce slow rolling. Watch the blanking bar as it moves up or down on the picture screen. The blanking bar should remain constant in vertical height at the bottom, middle, and top of the screen.

Vertical distortion can be caused by improper adjustment of controls as well as by defective circuits. Therefore, the first step in troubleshooting this symptom is to adjust the vertical linearity control. There is no problem if vertical linearity can be obtained with the control at mid range. If the vertical linearity control must be on full (or near full), there is a problem in the vertical sweep circuits.

One problem here is the interaction of controls. For example, if the sawtooth sweep is low in amplitude (due to a defect not associated with nonlinearity), the vertical height control or the vertical output bias control may be advanced to get proper vertical height. These extremes in circuit resistance may make it impossible to get a linear picture, no matter how the linearity control is adjusted.

If the problem cannot be cured by adjustment, the next step is to localize the fault. Start by measuring all waveforms and voltages in the vertical sweep circuits. Keep in mind that the vertical sweep will not be linear if the sawtooth waveform is not linear. Check waveform linearity, starting with the vertical output transistor and working back to the vertical oscillator.

6-4.2.5. Line Splitting

The problem of line splitting (line pairing or poor interlace) is often associated with vertical sweep circuits. Although the trouble does appear in the vertical sweep circuits, the actual cause is usually in related circuits. For example, the trouble can be caused by an open capacitor in the integrator portion of the sync separator or by horizontal pulses leaking back into the vertical sweep.

If you suspect that the horizontal pulses are present in the vertical sweep, observe the vertical sweep output waveform with the oscilloscope retrace blanking function disabled. If there are any horizontal pulses mixed with the vertical sweep, they will appear as a pulse train on some portion of the vertical sweep output waveform. An example of this is shown in Fig. 6–11.

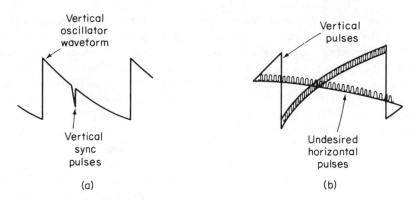

Figure 6–11. Waveforms in vertical sweep circuits.

If the horizontal pulses are present, the most likely causes are to be found in the horizontal circuits, such as corona discharge from the high-voltage rectifier, leakage between leads in the two circuits (not too common in printed circuit sets), and breakdown in the high-voltage section (resulting in a spark discharge being picked up by the vertical circuits).

If there are no horizontal pulses in the vertical circuits but there is definite line splitting, the problem is in the integrator portion of the sync circuit, with an open integrator capacitor the likely suspect. Generally, when line splitting is constant, the problem is in the integrator or is the result of leads being too close together. When splitting is intermittent, the cause is usually associated with the high-voltage section (spark discharge, radiation from high-voltage lead, etc.).

6–5. SYNC SEPARATOR CIRCUITS

The sync separator circuits function to remove the vertical and horizontal sync pulses from the video circuits, and apply the pulses to the vertical and horizontal sweep circuits, respectively. The sync separator circuits also function as clippers and/or limiters to remove the video (picture) signal and any noise. Thus, the sweep circuits receive sync pulses only and are free of noise and signal (in a properly functioning set).

Sync separator circuits are not standardized in solid-state TV. For example, to get the desired clipper/limiter action, the input transistor of one sync separator is reverse-biased (class C). Only the peaks of the sync pulses turn the transistor on and appear at the collector. In another solid-state circuit, the input transistor is zero-biased (class B) or slightly forward-biased, so that the sync pulses drive the transistor into saturation at a level well above the signal and/or noise level. In other solid-state sync separator circuits, two stages are used, one with reverse bias and the other with zero-bias (or slight foward bias). Thus, both clipping and limiting are obtained.

No matter what bias system is used, the vertical sync pulses are applied through a capacitor/resistor low-pass filter (often called the vertical integrator) to the input of the vertical sweep circuit (Sec. 6–4). The horizontal sync pulses are applied to the AFC section of the horizontal oscillator where they are compared with the horizontal sweep as to frequency and phase (Sec. 6–3).

Figure 6–12 is the schematic diagram of a typical sync separator circuit. Note that the video input is negative-going and that the input transistor Q_1 is PNP. Thus, both the sync pulses·and the signal/noise turn Q_1 on. However, with Q_1 biased near zero, the large sync pulses drive Q_1 into saturation at a level never reached by the signal and/or noise. The second sync separator transistor Q_2 is reverse-biased, so that the first portion (about 0.3 V) of the Q_1 output is clipped. This further removes any signal and/or noise.

The output of Q_2 is applied to the low-pass filter (vertical integrator) of C_6 and R_{11}. The 60–Hz vertical signals are applied to the vertical sweep circuit input (the primary of the vertical oscillator transformer) through series and shunt diodes CR_1 and CR_2. These diodes are arranged to pass the sync pulses to the vertical sweep input but prevent the vertical oscillator pulses from passing back into the sync circuits.

The output of Q_2 is also applied to Q_3 which acts as a phase splitter or phase inverter. The output from Q_3 is applied to the horizontal AFC circuits. The phase inverter is required when the horizontal AFC circuits are of the balanced type (Fig. 6–7).

6–5.1. Recommended Troubleshooting Approach

Failure of the sync separator circuits can produce symptoms similar to those produced by failure in other circuits. For example, if the low-pass filter (vertical integrator) circuit fails, there will be a loss of vertical sync or poor vertical sync. This same symptom can be produced by failure of the vertical sweep circuits as well. Also, operation of the sync separator circuits depends on signals from other circuits. For example, if the video output is low, the sync separator output will be low or possibly absent.

Because of these conditions, the only practical troubleshooting approach for the sync separator circuits is to measure waveforms at all outputs and inputs, followed by d-c voltage measurements at all transistor elements.

However, before going inside the set, the first logical step is to try correcting the problem by adjustment of horizontal and/or vertical controls. If the problem cannot be eliminated by adjustment of controls, then make the waveform and voltage measurements. Keep in mind that if it is necessary to set a control to an extreme (at one end or the other) or if the control must be reset repeatedly, look for a marginal component failure.

6–5.2. Typical Troubles

The following paragraphs discuss symptoms that could be caused by defects in the sync separator circuits.

6-5.2.1. No Sync

If horizontal and vertical sync is absent, the first step is to check for proper sync pulses at the input of the sync separator (from the video amplifier). If these pulses are not normal (particularly if they are low in amplitude), the problem is ahead of the sync separator. For example, if the tuner, IF amplifiers, or video amplifiers have marginal defects (low gain, improper alignment, etc.), the sync pulses to the sync separator will be abnormal. Thus, both the horizontal and vertical sync outputs from the separator will be abnormal.

If the sync pulses from the video amplifier are good, the next step is to check the sync pulses at the last point common to both the horizontal and vertical sync. In the circuit of Fig. 6–12, such a point is the collector of Q_2. If the pulses at the Q_2 collector are not good with a good input from the video amplifier, the problem is in Q_1 or Q_2 (or the related parts). Further waveform measurements (between Q_1 and Q_2) and/or voltage measurements can be used to isolate the problem. The most likely causes of trouble in the circuits of Q_1 and Q_2 are open or leaking coupling capacitors, leaking transistors, and leaking bypass capacitors.

6-5.2.2. No Vertical Sync

If vertical sync is absent or critical (requires frequent adjustment and will not hold), but there is good horizontal sync, the problem is in the vertical integrator or the vertical sweep circuits. Therefore, the first step is to measure the waveform at the vertical integrator output. In Fig. 6–12, such a point is the primary of transformer T_1. If the pulses are good at this point, the problem is in the vertical sweep circuits (refer to Sec. 6–4). If the pulses are abnormal at the input to the vertical sweep, the problem is probably in the integrator.

When measuring waveforms of vertical and horizontal sync pulses, it is convenient to set the oscilloscope sweep to 30 Hz (for vertical) and 7875 Hz (for horizontal). This will display two cycles of the corresponding pulses.

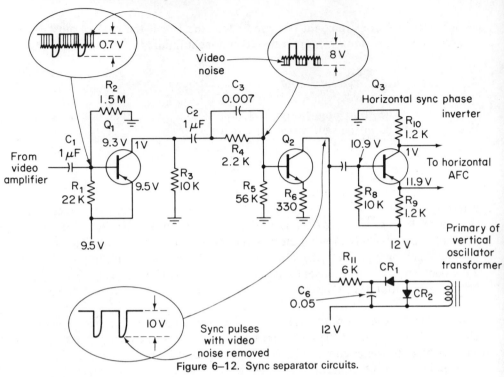

Figure 6–12. Sync separator circuits.

Some oscilloscopes designed specifically for TV service are provided with these sweep rates.

When measuring vertical pulses (sweep rate at 30 Hz) after the vertical integrator, there should be no horizontal pulses present (since the vertical integrator acts as a low-pass filter). However, when checking vertical pulses ahead of the integrator (say, at Q_1 or Q_2 in Fig. 6–12), the horizontal pulses may appear on the oscilloscope display. When the oscilloscope is set to measure horizontal pulses (set at 7875 Hz), the vertical sync pulses should not appear since they are so slow in relation to the horizontal sync pulse frequency.

When measuring ahead of the vertical integrator, which is the last point common to both horizontal and vertical sync, look for any noise or video signal that may have leaked through. If there is any noise or video at the output of Q_2, this is usually a sign of improper bias on Q_1 and/or Q_2. With improper bias, the undesired noise and video may not be removed by the normal clipping or limiting action. Of course, improper bias should show up as an abnormal voltage.

When measuring the output of the vertical integrator, look for a kickback voltage from the vertical sweep circuit. Such kickback produces a display similar to that shown in Fig. 6–13. As discussed in Sec. 6–4, the vertical oscillator uses feedback between secondary windings to sustain oscillation. The large surge of current through the secondary winding is reflected back to the primary and possibly into the sync separator circuits. The purpose of CR_1 and CR_2 is to prevent the kickback from entering the sync circuits. Note that not all solid-state circuits have diodes such as CR_1 and CR_2. However, there is usually some similar circuit.

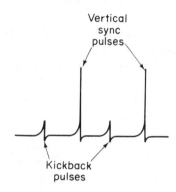

Figure 6–13. Kickback pulses present in sync separator circuits.

6-5.2.3. No Horizontal Sync

If horizontal sync is absent or abnormal with good vertical sync, the problem is in the horizontal portion of sync separator or the horizontal sweep circuits. Therefore, the first step is to measure the waveform at the horizontal portion of the separator. In Fig. 6–12, such points can be the collector and emitter of Q_3. The waveforms should be identical in this case since Q_3 functions as a phase inverter and provides two pulses (180° out of phase) to a balanced horizontal oscillator AFC circuit (Sec. 6–3). An unbalanced AFC requires only one horizontal pulse input from the sync separator.

If the pulses are good at the input of the AFC circuit, the problem is in the horizontal circuits (refer to Secs. 6–2 and 6–3). If the pulses are abnormal at the AFC input, the problem is in the horizontal output portion of the sync separator.

Keep in mind that horizontal sync troubles can be the result of poor high-frequency response, while vertical sync problems are generally caused by poor low-frequency response. Of course, this does not apply to sync problems caused by failure of a specific component.

6-5.2.4. Picture Pulling

There are several forms of picture pulling common in solid-state sets. Often, the nature of the picture pulling symptom can pinpoint the trouble. For example, when picture pulling appears to be steady and there is a bend in the image, this indicates that the horizontal AFC circuits are receiving distorted pulses. A likely cause is vertical kickback pulses entering the sync separator and distorting the horizontal output. If this condition is suspected, check for vertical kickback pulses in the sync separator and for distortion of the horizontal pulses with an oscilloscope.

If the picture pulling appears unsteady and particularly if the pulling tends to follow the camera signal, this indicates poor sync separation. That is, the video output is not being clipped and limited sufficiently to remove all camera signals from the circuits. If this condition is suspected, check transistor bias, particularly the bias of Q_1 and Q_2 in Fig. 6–12 (or their equivalents) in the circuit being serviced. This bias sets the signal levels at which clipping and limiting occur. In general, if bias is too high, the camera signal will be cut out but the sync signals will be attenuated. If bias is too low, the sync signals will be good but some camera signal can also pass, resulting in picture pulling, distortion, and so on.

Once the picture-pulling symptoms have been analyzed, the logical suspicion should be confirmed with waveform and voltage measurements. Start with a waveform measurement of the horizontal pulses to the horizontal AFC input. If these waveforms show a steady distortion, look for vertical kickback. If the waveforms show a fuzzy baseline, look for poor sync separation and the presence of camera signal. For example, in Fig. 6–12 there may be a camera signal at the output of Q_1, but there should be none at the output of Q_2.

Do not confuse picture pulling with distortion of the raster. If the raster is bent, the problem is not in the sync separator circuits but in the horizontal or vertical sweep circuits. If the raster edges are sharp but the picture is pulling or bent, then the problem can be in either the sweep or separator circuits. Check the raster alone by switching the channel selector to an unused channel.

6–6. RF TUNER CIRCUITS

In tuner troubleshooting some technicians will only replace the tubes and transistors and possibly check alignment. They prefer to send defective tuners to specialized repair shops. However, since transistors are more difficult to check by substitution than tubes, further troubleshooting is required.

All of the time-tested methods for troubleshooting tube-type tuners apply to transistor tuners, but possibly with slight modification. In any tuner, if the problem (weak picture, loss of picture, etc.) is observed on only one channel, the trouble is in the tuner. Troubleshooing should start with the circuits (coils and capacitors) that apply to the defective channel. If a problem appears on all channels, the trouble can be in the tuner or in the IF stages. Thus, the first step is to localize the problem.

Figure 6–14 is the schematic diagram of a typical RF tuner. Note that the tuning coils are of the turret type, where a separate set of drum-mounted coils is used for each channel and the entire drum rotates when the channel is selected. Most transistor TV tuners are of the turret type, with a very small percentage of the switch type in use. (Switch type tuners have series-connected coils mounted on wafer switches.)

All three stages of the tuner (RF amplifier, oscillator, and mixer) are connected in the common-emitter (CE) configuration. This is typical for transistor tuners, although a few tuners use the common-base (CB) configuration. The RF amplifier Q_{301} is neutralized by C_{308}. Neutralization is generally required for all solid-state RF amplifiers used in tuners. If the neutralization capacitor is open, the RF amplifier can break into oscillation.

The RF amplifier base is connected to the AGC network. Most solid-state TV AGC networks operate on a different basis from vacuum-tube systems. With solid-state, a strong signal increases the forward bias and drives the transistor into the saturation region of collector current, thus reducing gain.

Most solid-state tuners are provided with at least one test point (often called the "looker" point). In the circuit of Fig. 6–14, the test point is at the collector of Q_{303}. This test point is used mostly to monitor the tuner output with an oscilloscope. The same point can also be used to inject a signal into the IF amplifiers.

6–6.1. Recommended Troubleshooting Approach

It is often difficult to decide if the tuner or the IF amplifiers are at fault. The same symptoms can be produced by problems in either section of the set. For example, if the tuner RF amplifier gain is low, the same basic symptoms (weak picture and sound, poor contrast, etc.) will be produced as when the IF amplifier gain is low.

Some technicians prefer to make a quick check by shorting the RF amplifier AGC point (base of Q_{301}) to ground. This will remove the forward bias on the RF amplifier and usually cut the stage off. If this produces considerable change in the picture, the RF amplifier is probably good. If little change is noticed when the RF amplifier is cut off, the RF amplifier is *probably* at fault.

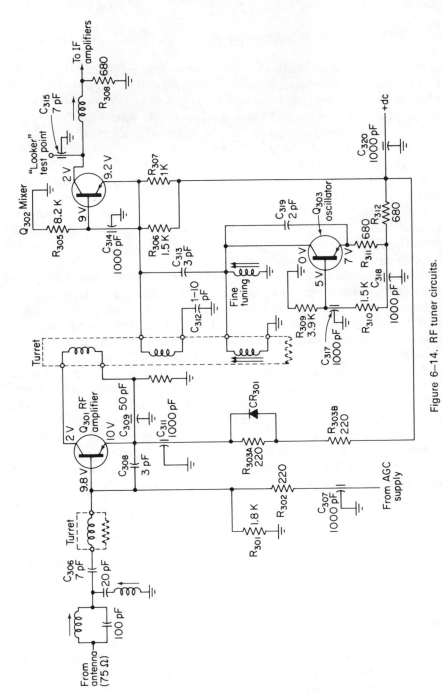

Figure 6–14. RF tuner circuits.

If adequate test equipment is available, the tuner should be checked thoroughly before replacing it or sending it to an outside shop. The best method is to apply a sweep frequency signal (with markers) to the antenna input and monitor the tuner output at the mixer test point with an oscilloscope. The same procedures, described in the author's *Handbook of Oscilloscopes. Theory and Application,* Prentice-Hall, Inc., Englewood Cliffs, N.J. 1968, can be used for tuner alignment. Always refer to the manufacturer's service literature for detailed alignment procedures. If the tuner output is good, as measured at the tuner looker point, the problem is likely to be in the IF stages.

If the tuner output is not good, the next step is isolation of the fault to one of the three tuner stages. One simple approach is to apply an RF signal (modulated by an AF tone) to various points in the tuner and observe the picture display. Tune the RF signal generator to the channel frequency. If the tuner is operating normally, a series of horizontal bars will appear on the picture-tube screen. The number of bars depends on the frequency of the AF modulating tone.

If the picture display is normal with the RF signal injected at the collector of Q_{301} but not normal with a signal at the antenna, the problem is likely to be in the RF amplifier Q_{301}. Next, inject the signal at the base of Q_{301}. If the picture display is normal here, the problem is in the network between the antenna and the base of Q_{301}.

If it is not possible to get a signal through the tuner at any point, change the RF generator frequency to the frequency used by the IF amplifier. If the IF signal passes (produces a display on the picture tube), the most likely trouble is a defective oscillator Q_{303}.

6–6.2. Typical Troubles

The following paragraphs discuss symptoms that could be caused by defects in the RF tuner circuits.

6 6.2.1. No Picture or Sound

If a raster is present but there is no picture or sound, the first step is to inject an IF signal at the looker point or monitor the tuner output at the same point, whichever is most convenient. If the fault is definitely isolated to the tuner, the next step is to check the raster pattern for snow. If there is a complete absence of snow, the defect is likely to be in the mixer stage rather than in the RF amplifier. If the mixer stage and IF amplifiers are good, there will be some snow even if the RF amplifier is completely dead.

If it appears that either the RF amplifier or mixer is completely dead, measure voltages at the transistor elements, and/or inject RF signals at both stages as described in Sec. 6–6.1. Note that both the RF amplifier and mixer are forward-biased when operating normally. Keep in mind that the RF amplifier is usually forward-biased by the AGC network. If this forward bias is missing or abnormal, the problem can be in the AGC circuits rather than in the tuner. A simple test is to apply the required forward-bias voltage to the AGC line where it enters the tuner. If this restores normal operation, check the AGC circuits for a defect. If the service literature is available, check for the correct amount of forward bias from the AGC circuit to the RF amplifier under no-signal conditions. Usually, the service literature will show both the no-signal and full-signal bias values. Use the no-signal bias value to make the test.

If an RF signal will not pass but an IF signal will pass as described in Sec. 6–6.1, the oscillator is suspect. As a quick test try injecting a signal at the oscillator frequency, with the set tuned to a normal, strong channel. Use an unmodulated RF signal generator. Preferably, the signal should be injected at the same point as is the tuner oscillator. In the circuit of Fig. 6–14, the oscillator signal is injected at the emitter of Q_{302} through C_{313}. If operation is restored when the external oscillator signal is injected, the tuner oscillator circuit is defective.

6-6.2.2. Poor Picture or Sound

The basic procedure for troubleshooting a poor picture and sound symptom (snow, weak sound, poor contrast, etc.) is the same as for no sound or picture. That is, the trouble must be isolated to the RF tuner and then to a particular stage in the tuner. However, it is necessary to have a greater knowledge of the circuit's capabilities to troubleshoot a poor performance sympton. For example, if the tuner oscillator is completely dead, operation can be restored by injecting a signal at the oscillator frequency. This pinpoints trouble to the tuner oscillator. However, if the tuner oscillator is producing a weak signal, an injection test may prove confusing. Ideally, the service literature should be consulted to find the normal amplitude of the oscillator output.

6-6.2.3. Hum Bars or Hum Distortion

In solid-state TV sets, the presence of hum bars on the picture, hum distortion, or poor sync accompanied by hum symptoms is generally the result of a failure in the power supply. In vacuum-tube sets, hum is often caused by cathode-heater leakage or some form of leakage in the tube elements. This is not the case in solid-state sets. If there is any evidence of hum in the display, check the d-c input to the RF tuner for the presence of a 60- or 120-Hz hum. If 60- or 120-Hz signals are present on any of the d-c lines, check the power supply as described in Sec. 6–1. Most solid-state sets now use full-wave rectifiers in the low-voltage power supply. Therefore, hum will appear as 120-Hz signals on the d-c line if the power supply filter or regulator fails.

6-6.2.4. Picture Smearing and Sound Separated from Picture

When the picture appears to be smeared, with the trailing edges of images not sharply defined, the trouble is usually the result of poor response and is often accompanied by a separation of picture and sound. That is, when the fine tuning is adjusted for the best picture, the sound is poor, and vice versa. This symptom can be caused by poor alignment or problems in the AGC system.

Likewise, the problem can be in the IF stages. Thus, the first step is to localize the problem to the RF tuner with a response test as described in Sec. 6–6.1. Next, apply a fixed d-c bias to the AGC line equal to the no-signal bias. (This is often known as *clamping* the AGC line.) Keep in mind that virtually all solid-state TV AGC systems apply a forward bias to the RF tuner amplifier under both no-signal and full-signal conditions.

If the problem is eliminated with the proper forward bias applied, look for trouble in the AGC network, as described in later sections. If the problem remains with correct bias applied, check the response pattern of the tuner. If the response pattern is not good, try correcting the condition by alignment. Always follow the service literature instructions for alignment. Generally, the AGC line must be clamped during tuner alignment. If it is not spelled out by the service literature, clamping the AGC line with a forward bias is usually implied.

Many solid-state tuners are provided with a delayed AGC function. This is accomplished by diode CR_{301} in Fig. 6–14. CR_{301} requires about 0.5 V to conduct. If the signal increases to a point where the drop across R_{303A} is more than 0.5 V, CR_{301} will conduct and short R_{303A}. This will place a large forward bias on Q_{301} and drive the transistor into saturation, thus reducing the gain.

Whenever tuner response is poor, with the AGC network and IF stages cleared, and the problem cannot be corrected by alignment, look for such defects as open neutralizing capacitors, open bypass capacitors, and loose or poorly grounded tuner shields. Any of these most likely defects can cause poor tuner operation but will not show up as incorrect d-c voltages at the transistor elements.

6-6.2.5. Ghosts

If there are ghosts (double images or repeats) in the picture display, it is possible that the tuner is at fault. Of course, ghosts can be caused by propagation problems (where the TV transmission is reflected from a building or similar object, and the reflected signals arrive at a slightly different time than the nonreflected signals) or by antenna problems. Therefore, the first step is to isolate the problem to the set rather than to the antenna or outside conditions. Generally, when ghosts are caused by reflections, the ghosts will change from channel to channel or will sometimes disappear on some channels.

A more positive test is to apply a signal from a pattern-type generator to the tuner input. If the ghosts disappear, they are the result of outside causes. If the ghosts remain with a signal applied to the tuner, they are definitely circuit ghosts and can be caused by problems in the tuner. Generally, circuit ghosts are the result of problems in the IF stages or video stages but can be in the tuner.

When there are circuit ghosts in the tuner, it is generally certain that the tuner has a very sharp response due to an abnormally high Q in one of the tuned circuits. This will show up as very sharp peaks in the tuner response curve. The most likely causes, other than poor alignment, are open capacitors, particularly bypass capacitors or the RF amplifier neutralization capacitor.

6-6.2.6. Picture Pulling

As discussed in previous sections, picture pulling is usually a problem in the sweep or sync separator circuits. However, the problem can also be caused by the tuner. As a first test, check the edges of the raster. If they are sharp and well defined, the sweep circuits are probably good. If an analyzer-type generator is available, inject an IF signal (complete with sync pulse) at the input of the IF stages (at the tuner looker point if convenient). If the pulling symptom disappears, the problem is in the tuner. If the pulling symptom remains, the problem is in the IF stages, video stages, or most likely in the sync separator.

If the problem is definitely localized to the tuner, proceed with the localization and isolation steps discussed previously: monitor the tuner response pattern at the looker point, measure the waveforms and voltages at each stage, try to correct the poor response pattern by alignment, and check the response pattern with the AGC line clamped and unclamped. One or more of these steps should pinpoint the problem.

6-6.2.7. Intermittent Problems

When both the picture and sound are intermittent, the tuner is usually suspected but the actual cause can be in another circuit. For example, if the AGC circuit is defective and intermittently removes the forward bias to the RF amplifier transistor, the RF amplifier can be cut off, removing both sound and picture. At best, intermittent problems are difficult to locate.

In the case of a tuner, the best bet is to monitor the circuits and watch for changes when the intermittent occurs.

As a minimum, monitor the tuner output at the looker point, the AGC line, and the d-c voltage line into the tuner. If either the d-c voltage or the AGC line voltage changes considerably when the intermittent condition occurs, the problem is likely not to be in the tuner but in the power supply or AGC circuits. If neither the d-c voltage nor the AGC line voltage changes but the tuner output display is affected during the intermittent condition, the problem is in the tuner. The most likely causes are cold solder joints, breaks in printed circuit wiring, and intermittent capacitors (particularly in the RF amplifier and mixer stages). In rare cases, the transistor will become intermittent. If all else has failed to localize the trouble, try applications of heat and cold to the tuner transistors, as described in Chapter 2, or simply try substitution of the transistors, whichever is more convenient.

6-6.2.8. UHF Tuner Problems

Figure 6–15 is the schematic of a solid-state UHF tuner. Note that there is no RF amplifier as such, that a diode CR_1 is used as the mixer,

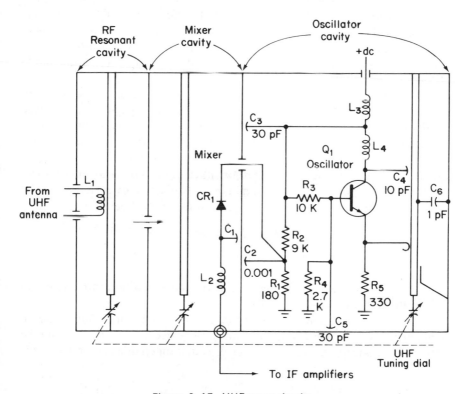

Figure 6–15. UHF tuner circuits.

and that a transistor is used as the oscillator. Also note that the tank circuits are in the form of resonant cavities, tuned by variable capacitors connected to the UHF fine-tuning knob. No AGC is provided for the UHF tuner.

All the problems that apply to a vacuum-tube UHF tuner apply to the solid-state version. For example, if the physical structure of the resonant cavities is changed, the resonant frequency will change. That is, if the tuner shields are removed, bent, or otherswie distorted or if shield screws are loose, the resonant frequency can change. A minor frequency change will make the UHF fine-tuning knob markings incorrect. A major frequency change can make the UHF tuner totally inoperative.

Ideally, the UHF tuner should be checked by means of a UHF signal at the antenna input (preferably from a UHF sweep generator with markers) and an oscilloscope at the IF output. In this way, the tuner gain, bandwidth, and overall frequency response can be checked quickly and accurately. Unfortunately, many shops are not equipped with a UHF generator.

The first step in troubleshooting is to check operation of the set on a UHF channel with a known good signal. If the UHF reception is dead or very poor, with a known good signal and with good VHF reception, the problem is in the tuner.

The next step is to measure all the voltages at the transistor elements. Note that Q_1 is forward-biased. Also note that the mixer diode CR_1 has a bias voltage placed on its cathode. If all voltages appear normal, the next step is to check the mixer diode CR_1 and oscillator transistor Q_1 by substitution. The front-to-back ratio of CR_1 can be checked if desired. However, if the diode appears to be defective, it must be substituted.

If none of these procedures localize or cure the problem, look for open capacitors in the tuner. Usually, a shorted or leaking capacitor will produce an abnormal voltage at one or more of the circuit elements, where an open capacitor will leave the voltages normal but make the circuit operate improperly. There are exceptions of course. For example, if capacitor C_6 in Fig. 6–15 is shorted or leaking, the resonant cavity will be seriously detuned, possibly to the point where Q_1 might not oscillate. However, the circuit voltages will remain virtually unchanged from normal operation.

6–7. IF AND VIDEO DETECTOR CIRCUITS

The basic functions of the IF and video detector circuits in solid-state sets are the same as for vacuum-tube sets. That is, the circuits amplify both picture and sound signals from the RF tuner mixer circuit, demodulate both signals for application to the video amplifier and sound IF amplifiers, and trap (or reject) signals from adjacent channels. Therefore, the basic troubleshooting methods for tube-type IF circuits can be applied to solid-state circuits.

Figure 6–16 is the schematic diagram of a typical IF and video detector circuit. Three stages of amplification are used, with each stage in the common-emitter configuration. All stages are forward-biased. The first two stages Q_1 and Q_2 receive their forward bias from the AGC circuit. This same AGC line is connected to the RF amplifier in the tuner, as discussed in Sec.

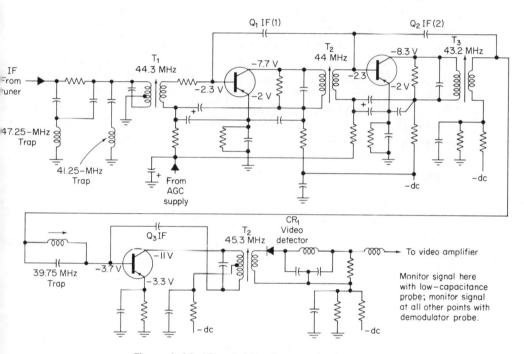

Figure 6–16. IF and video detector circuits.

6–6. On strong signals, the forward bias is increased, driving Q_1 and Q_2 into saturation and reducing gain. The third IF stage Q_3 does not have any AGC control. All three stages are neutralized to prevent oscillation in the IF circuits. If any of the neutralizing capacitors are open, the corresponding stage can break into oscillation.

Each stage is tuned at the input and output by corresponding transformers (T_1 through T_4). The stages are stagger tuned; that is, each transformer is tuned to a different peak frequency. This gives the overall IF amplifier circuit a bandwidth of about 3.25 MHz.

Three traps are used. The 41.25- and 47.25-MHz traps are series resonant, while the 39.75-MHz trap is parallel resonant. As is usual, the traps are adjusted for a minimum output signal at the video detector, when a signal of the corresponding frequency is injected at the IF input.

There are no test points as such provided on most solid-state IF circuits. However, input from the tuner is usually applied through a coaxial cable. This cable can be disconnected and used as a signal injection point for overall IF amplifier tests and alignment. In some cases, the looker point on the RF tuner can be used instead. Output of the IF stages can be measured at the video detector CR_1. The video output is typically about 1 V for most solid-state sets. That is, the peaks of the sync pulses are about 1 V, with the picture signals about 0.25 to 0.5 V. Most solid-state IF circuits are mounted on printed circuit boards so that the transistor base elements are accessible. Signals can be injected at any of the bases to trace performance through the IF stages.

In some solid-state sets, the entire IF function is performed by an IC. In such cases, the usual rules for IC troubleshooting apply. That is, a signal can be injected at the input and the results are monitored at the output. Supply voltages to the IC can also be checked. If there is any defect in the IC, the entire IC must be replaced as described in Chapter 2.

6–7.1. Recommended Troubleshooting Approach

The recommended approach for troubleshooting IF stages depends largely on the available test equipment. Ideally, an analyzer-type generator can be used. These generators duplicate the signals normally found at the mixer output and the video detector output (as well as several other signals). If the picture display is good with a signal injected at the video detector output (video amplifier input) but not good with a signal injected at the IF input (coaxial cable from the mixer), the problem is in the IF amplifiers. A possible exception is with defective AGC circuits. This problem can be eliminated by clamping the AGC line with the appropriate voltage. If the problem remains with the correct AGC bias voltage applied from a fixed external source, the problem is in the IF stages.

If an analyzer-type generator is not available, the next recommended test equipment setup is a sweep generator (with markers) and an oscilloscope. (Many technicians prefer this to the analyzer.) The sweep generator signal is injected at the IF input (cable from mixer), and the signal is monitored at various points throughout the IF stages with the oscilloscope. A demodulator probe is required for the oscilloscope if individual IF stages are to be monitored. A basic low-capacitance probe can be used if the output of the video detector is monitored. Keep in mind that signals through the IF stages of a solid-state TV are quite low (less than 1 V). Therefore, the oscilloscope must have considerable vertical gain. Operate the horizontal sweep of the oscilloscope at 30 Hz so that two cycles of sync pulses are displayed.

If analyzer and sweep generators are not available, use an RF generator (modulated by an AF tone). Tune the RF generator to the approximate center frequency of the IF amplifiers and inject the signal at the IF amplifier input (cable from mixer) and at the bases of each transistor in turn. If the IF amplifier is operating normally, a series of horizontal bars will appear on the picture-tube screen. (The number of bars depends on modulation frequency.)

6–7.2. Typical Troubles

The following paragraphs discuss symptoms that could be caused by defects in the IF amplifier stages.

6-7.2.1. No Picture or Sound

If a raster is present but there is no picture, no sound, or the sound is very weak and noisy, first inject an IF signal at the IF amplifier input. If operation is normal, the problem is in the tuner rather than in the IF stages. If the fault appears to be isolated to the IF stages, then clamp the AGC line and repeat the injection test. If the problem is cleared, look for trouble in the AGC circuits (as described in later sections).

If the fault is definitely isolated to the IF stages, inject an IF signal at the base of each stage as described in Sec. 6–7.1. Then measure the voltages at all transistor elements. This should isolate any problem serious enough to cause a no-picture/no-sound symptom.

6-7.2.2. Poor Picture or Sound

The basic procedure for troubleshooting a poor picture and sound symptom (snow, weak sound, poor contrast, etc.) is the same as for no sound or picture. That is, the trouble must be isolated to the IF circuits, and then to a particular stage. This may not be as easy to isolate as when there is complete failure of a stage. For example, the service literature rarely gives the gain per stage for the IF circuits. However, there should be some gain for each stage.

Generally, a poor picture and sound symptom is the result of low gain in one or more stages. Most of the troubles that cause low gain will show up as an abnormal voltage. For example, a leaking or shorted capacitor or an open transformer winding will produce at least an abnormal voltage at one or more transistor elements. Even if a transistor has low gain, the collector and/or emitter voltages will be abnormal. (Generally, the emitter will be low and the collector will be high.)

Open capacitors can sometimes cause low gain, without substantially affecting voltages. For example, if a capacitor across an interstage transformer winding opens, the transformer resonant circuit will be affected. If the transformer is seriously detuned, gain will be reduced. If an emitter bypass capacitor opens, the a-c gain of a stage will be reduced drastically, while the transistor voltages will remain substantially the same.

6-7.2.3. Hum Bars or Hum Distortion

Unlike vacuum-tube sets where hum problems (hum distortion, hum bars in picture, and poor sync plus hum) can be caused by cathode-heater leakage in one of the tubes, hum in solid-state sets is generally the fault of the power supply filter. One possible exception is an open decoupling capacitor in one of the IF stages. Since the output of the IF circuit is fed into the video amplifier where vertical blanking pulses are applied (Sec. 6–8), these pulses can enter the IF stages if a decoupling capacitor is open. As is the case with any suspected open capacitor, try connecting a known good capacitor in parallel with the suspected capacitor.

6-7.2.4. Picture Smearing, Pulling, or Overloading

Generally, these symptoms are associated with the video amplifier section rather than with the IF, particularly when sound is good. However, if the IF stages are improperly aligned or if there is a defective component, making proper IF alignment impossible, the same symptoms will occur even though the sound channel will pass.

To eliminate doubt, inject a video frequency signal at the input of the video amplifier (output of video detector). If the picture problems are eliminated, the trouble is in the IF stages. Next, clamp the IF stages to see if the problem is eliminated. If the symptom disappears with the correct bias applied, look for trouble in the AGC network, as described in later sections.

If the problem remains with correct fixed bias, check the response pattern of the IF stages, following the manufacturer's service literature. If the response pattern is not good, try correcting the condition by alignment. If any particular stage cannot be aligned, look for such defects as open capacitors, poorly grounded IF coil shields, and open damping resistors.

6-7.2.5. Intermittent Problems

When both the picture and sound are intermittent, either the tuner or IF amplifiers is usually suspected. However, the AGC circuit or the low-voltage power supply could be at fault. In any case first the problem should be localized to the IF stages by monitoring the tuner as described in Sec. 6–6.2.7. If the intermittent condition occurs but the tuner indications are good, the problem is in the IF stages.

Next, monitor the IF stages and watch for changes when the intermittent occurs. As a minimum, monitor the video detector output, the AGC line, and the d-c voltage line of the IF circuits. In solid-state sets, the IF circuits are often mounted on a separate printed circuit board with one d-c voltage distribution point. The video detector output can be monitored by an oscilloscope with a low-capacitance probe. Any other point in the IF circuits (ahead of the video detector) will require a demodulator probe for monitoring.

If neither the d-c voltage nor the AGC line voltage changes but the video detector output (or other point being monitored) is affected during the intermittent condition, the problem is in the IF amplifier stages. As in the case of any solid-state intermittent, the most likely causes are cold solder joints, breaks in printed circuit wiring, and intermittent capacitors. Intermittent transistors are rare. If all else has failed to localize the trouble, try applications of heat and cold to the IF transistors, as described in Chapter 2, or try transistor substitution.

6–8. VIDEO AMPLIFIER AND PICTURE-TUBE CIRCUITS

Video amplifier circuits have several functions, and there are many configurations used in solid-state sets. Thus, it is very difficult to present a typical troubleshooting approach. However, most video amplifier circuits have three inputs and three outputs which can be monitored. If the inputs are normal but one or more of the outputs is abnormal, the problem can be localized to the video amplifier circuits.

Figure 6–17 is the schematic diagram of a typical solid-state video amplifier circuit. Two stages, driver Q_1 and output Q_2, are used. Transistor Q_1 is connected as an emitter follower. Typically, the signal input and output at Q_1 are about 1 V. This signal consists of the sound, video, and sync pulses, as taken from the video detector.

The output of Q_1 is applied to the input of video output transistor Q_2, to

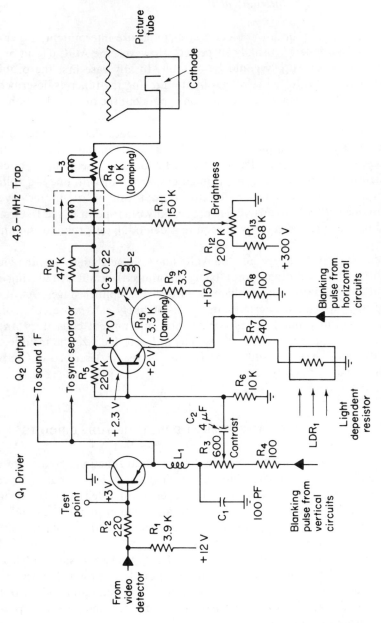

Figure 6–17. Video amplifier and picture-tube (video) circuits.

the input of the sync separator circuits (Sec. 6–5), and to the input of the sound IF circuits (Sec. 6–10). In the circuit of Fig. 6–17, the output of Q_1 is applied to Q_2 through contrast control R_3. In some solid-state circuits, the contrast control is part of the video output transistor circuit (typically in the emitter of the output transistor).

Video output transistor Q_2 is always a power transistor and is thus mounted on a heat sink. Transistor Q_2 amplifies the 1-V input to a level of about 25 to 50 V depending on the picture-tube screen size. Brightness control R_{12} sets the voltage level on the collector of Q_2. The output of Q_2 is applied through a 4.5-MHz sound trap to the cathode of the picture tube. This output contains the video information (picture signal) as well as the horizontal and vertical retrace blanking pulses. Since all of this information is applied to the cathode of the picture tube, the video (picture) signal is negative, while the blanking pulses are positive. In some solid-state picture-tube circuits, the blanking pulses are applied to the first control grid of the picture tube (and in rare cases to the filament) rather than being mixed with the video pulses. In still other cases, the mixed video and blanking pulses are applied to the control grid. Some solid-state picture-tube circuits do not have any horizontal blanking pulses. Figure 6–18 shows some typical solid-state picture-tube circuit configurations as well as typical voltages. Note that all of the voltages for operation of the picture tube (except for the filament) are supplied by the high-voltage portion of the horizontal output circuits (Sec. 6–2).

The vertical retrace blanking pulses from the vertical sweep circuits (Sec. 6–4) are applied to the emitter of Q_1. Horizontal blanking pulses are applied to the emitter of Q_2. Note that both Q_1 and Q_2 are forward-biased during normal operation. The bias on Q_2 is partially determined by light-dependent resistor LDR_1, connected between the emitter and ground. As the ambient lighting around the set varies, the bias on Q_2 (and thus the gain of the video amplifier) varies. This variable gain feature provides the necessary changes in picture contrast to accommodate changing conditions of ambient light.

6–8.1. Recommended Troubleshooting Approach

The recommended approach for troubleshooting video amplifier stages depends largely on the available test equipment. Ideally, an analyzer-type generator can be used. These generators duplicate the signals normally found at the video detector output. If the picture display is not good with a signal injected at the video amplifier input (video detector output), the problem is in the video amplifier or possibly in the picture tube and its circuits.

If an analyzer-type generator is not available, the next recommended test setup is a sweep generator (with markers) and an oscilloscope. The sweep generator signal is injected at the tuner input or the IF amplifier input, and the signal is monitored at the video detector output. If the overall response

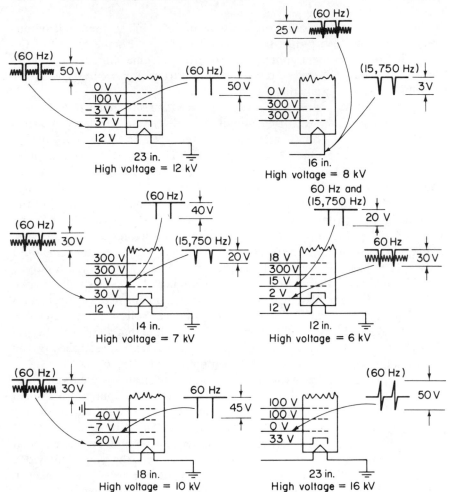

Figure 6–18. Typical picture-tube voltages and circuits.

through the RF (tuner) and IF stages is good but there are problems in the picture, the trouble is probably in the video amplifier.

As a quick check, the output of the video detector can be monitored with an oscilloscope (with the set tuned to an active TV channel). If there is a signal of about 1 V at the video amplifier input, any picture problems are probably in the video amplifier circuits.

Some technicians prefer to make a squarewave test of the video amplifier circuits. A squarewave of about 1 V is applied to the video amplifier input, and the resultant output is monitored on an oscilloscope at the picture-tube cathode. This checks the overall response of the video amplifiers, including the effect of the contrast and brightness controls. If squarewaves are passed without distortion, attenuation, ringing, and so on, the video amplifier circuits are operating properly.

6–8.2. Typical Troubles

The following paragraphs discuss symptoms that could be caused by defects in the video amplifier stages.

6–8.2.1. No Raster

A no-raster symptom is usually associated with failure of the power supply or horizontal sweep circuits rather than failure of the video amplifier. However, complete cutoff of the picture tube can be produced by failure in the video amplifier circuits, even though the high-voltage and picture-tube grid voltages are normal. For example, if brightness control R_{12} in Fig. 6–17 is open or making poor contact on the ground side, a high positive voltage is applied to the cathode of the picture tube. This will cut the picture tube off (no raster), regardless of signal conditions or other voltages.

If this condition is suspected, check all of the picture-tube voltages against the service literature. Pay particular attention to the filament and cathode voltages. A low filament voltage can cause the filament to glow but will not produce enough emission for a picture. If any one of the voltages is not normal, trace the particular circuit and look for such defects as shorted or leaking capacitors, breaks in printed wiring, and worn brightness controls. If all of the voltages appear normal, try substitution of the picture tube.

6-8.2.2. No Picture and No Sound

With a raster present, a no-picture, no-sound symptom is normally associated with failure of the tuner or IF stages rather than the video amplifier. However, in the circuit of Fig. 6–17, failure of Q_1 could cut off both sound and picture, even though good signals are available at the video detector output.

The first step is to monitor the signal at the base and emitter of Q_1. Note that a test point is provided at the base of Q_1. If the fault is definitely localized to Q_1 (good signal at base, absent or abnormal signal at emitter), the next step is to measure the voltages at all Q_1 elements. This should pinpoint the actual cause of trouble,

6-8.2.3. No Picture and Normal Sound

With a raster present and good sound, a no-picture symptom is definitely in the video amplifier circuits. First monitor the signal at the base and collector of Q_2 and at the cathode of the picture tube. This should be followed by measurement of the Q_2 element voltages. As in the case of any amplifier, total failure of the circuit is usually easy to locate. Look for such defects as an open coupling capacitor and coils, cold solder joints, open or worn contrast controls, breaks in printed circuit wiring, and the like.

Keep in mind that Q_2 is a power transistor and is therefore subject to being burned out if its heat sink is defective (bad thermal conduction between heat sink and transistor). Also, other circuit defects can cause Q_2 to burn out. For example, if coupling capacitor C_2 is shorted or badly leaking, the forward bias on Q_2 could cause excessive current flow.

6-8.2.4. Contrast Problems

Too little contrast (or weak picture) that cannot be corrected by adjustment of the contrast control is the result of poor gain or lack of signal strength. In the circuit of Fig. 6–17, the contrast control sets the signal level (or signal strength) applied to the video output transistor Q_2. The contrast control does not set the gain of either stage. This is typical for most solid-state video amplifier circuits, although there are a few solid-state circuits where the contrast control sets stage gain, as is typical for vacuum-tube sets. Poor picture contrast can be the result of a weak (low-emission) CRT filament. In this case, the CRT must be replaced, or if it is a color CRT, its filament may be rejuvenated. The latter procedure is not always successful.

If the symptom is poor contrast, the first step is to monitor the input to the video amplifier circuits (video detector output). If the voltage is about 1 V with low contrast, the problem is in the video amplifier circuits. The amount of gain provided by the video amplifier in normal operation depends, of course, on the circuit. In the circuit of Fig. 6–17, output transistor provides a gain of about 50. That is, the 1-V output from Q_1 is raised to about 50 V. In some solid-state video circuits (particularly those with three stages) there is an overall gain of about 100. That is, the video detector output is about 0.5 V, with 50 V applied to the picture tube. Keep in mind that these voltage and gain figures are typical for solid-state circuits. Always consult the service literature, when available.

If the gain is low or is suspected to be low, measure the transistor voltages and look for the usual problems associated with low gain: excessive collector-to-base leakage, leaking capacitors, low power supply voltages, worn controls, and so on.

Too much contrast (that cannot be corrected by adjustment of the contrast control) is the result of too much gain or excessive signal strength. The first step is to monitor the video detector output. If the signal voltage is between 1 V and 2 V with excessive contrast, the problem is in the video amplifiers.

If there is an excessive signal at the video detector output, look for trouble in the RF tuner or IF stages; if possible, consult the service literature for the correct video output signal level.

Too much contrast is often associated with a defect in the contrast control circuit. Therefore, the next step is to monitor all voltages associated with the contrast control. If the voltages appear to be in order, monitor the signal voltages at the input and output of the video circuits. Compare the signal voltage at the picture-tube cathode against the service literature. If gain is too high, look for some circuit condition that is biasing the output transistor Q_2 to an incorrect level. Solid-state amplifier circuits are usually designed on the basis of minimum transistor gain. If a particular transistor has more gain than normal, it is possible that a slight change in bias can cause excessive gain. A bias change can result from aging of parts and from defects in circuits connected to the video amplifiers. However, this will show up as an abnormal voltage.

Another possible cause of excessive contrast is a defective heat sink for the video output transistor. If the heat sink is not making good contact with the transistor case (for good thermal conduction), the transistor temperature can rise and increase transistor gain. If the rise is not too great, the transistor will not burn out, particularly if the set is used for short periods of time.

6-8.2.5. Sound in Picture

A sound-in-picture symptom (picture appears to be modulated by the sound) can be caused by problems in the RF tuner or IF stages and is often a result of poor alignment and improper fine-tuning adjustments. First localize the trouble by monitoring the output of the video detector with an oscilloscope and low-capacitance probe. If the video display appears to be modulated by sound (video display is fuzzy and/or varies with sound) the problem is ahead of the video amplifier stages. Keep in mind that both sound and video are present at the base and emitter of Q_1 in normal operation. However the sound should not modulate the video.

If the problem is definitely isolated to the video amplifier stages, first adjust the 4.5-MHz sound trap. In the circuit of Fig. 6–17, the sound trap is between the video output transistor and the picture-tube cathode. Any sound present at the collector of Q_2 should be attenuated fully when the sound trap is properly tuned.

If the problem can not be corrected by adjustment of the sound trap, look for defective components in the sound trap (open capacitor, shorted coil turns). Some solid-state video circuits do not have a sound trap as such. Instead, the sound is removed through a transformer tuned to 4.5 MHz. The transformer secondary passes the sound signal to the sound IF stages.

The transformer primary appears as a low-impedance short to 4.5-MHz-signals in the video circuits, thus preventing the signals from passing to the picture tube.

6-8.2.6. Poor Picture Quality

When picture quality is poor (lack of detail, smearing, circuit ghosts, poor definition, etc.), the video amplifiers are usually suspected, particulary if the sound is good. However, these same symptoms can be caused by problems in the tuner and/or IF stages. If an analyzer-type generator is available, the first step is to inject a video signal into the input of the video amplifiers and watch the display on the picture tube. If the trouble symptom remains, the problem is in the video amplifier circuits.

If an analyzer generator is not available, the frequency response of the video amplifiers can be checked using squarewaves. Response testing of solid-state video amplifiers is essentially the same as for vacuum-tube circuits. (The procedure is describe in the author's *Handbook of Oscilloscopes: Theory and Application,* Prentice-Hall, Inc., Englewood Cliffs, N.J., 1968.) In brief, squarewaves of approximately 1 V are injected at the video amplifier input, and the resultant display is monitored by an oscilloscope connected to the picture-tube cathode. A squarewave frequency of about 60 Hz is used initially. Then the frequency is increased to about 100 kHz. If the square-waves pass to the picture-tube cathode without distortion, frequency response of the video amplifier is good. If there is any distortion, the wave shape will provide a clue as to frequency response. For example, if the leading edge (left-hand side) of the squarewave is low or rounded, while the trailing edge is square, this indicates poor high-frequency response or excessive low-frequency response. If the leading edge has greater amplitude than the trailing edge, low-frequency response is poor. If there is overshoot or oscillation on the leading edge, there is rising high-frequency response (or "ringing"). Keep in mind that symptoms such as smearing or poor definition are usually the result of poor high-frequency response, while circuit ghosts are caused by rising high-frequency response.

From a practical standpoint, any frequency response problems are the result of changes in tuned circuits. Look for such problems as leaking capacitors that can change the resonant frequency of peaking coils (L_2 and L_3 in Fig. 6–17), solder splashes that have shorted out damping resistors across peaking coils, or in rare cases, shorted peaking coil turns.

About the only frequency response problem unique to solid-state video amplifiers is excessive junction capacitance. Transistor junctions have a capacitance value that can change with age or (more likely) temperature extremes. Such a change can upset the resonant frequency (and thus the frequency response) of associated circuits. This problem is not too frequent in original equipment but can occur when video amplifier transistors are replaced. Always use the correct replacement (an identical type if at all possible). Even with identical transistor replacements, it is possible that the junction capacitance can be incorrect (usually too much capacitance). Remember the old rule that a transistor is good only if it operates properly in circuit.

6-8.2.3. Retrace Lines in Picture

In the circuit of Fig. 6–17, both vertical and horizontal blanking pulses are fed into the video amplifier to blank the retrace lines. As discussed previously, some solid-state circuits do not provide blanking for the horizontal retrace. In other circuits, the blanking is applied to the picture-tube elements separately from the video amplifiers.

No matter what system is used, the obvious first step is to monitor the blanking pulses (with an oscilloscope and low-capacitance probe) at the point where they enter the video amplifier or at the picture-tube element. If the blanking pulses are absent, trace the particular circuit back to the pulse source.

If the blanking pulses are low in amplitude, the typical symptom is the presence of retrace lines only when the brightness control is turned up. If this is the case, check the pulse amplitude against the service literature.

6-8.2.3. Intermittent Problems

In a circuit such as Fig. 6–17, an intermittent condition in Q_1 or its related circuit elements will produce an intermittent sound and picture symptom. This same symptom can be produced by problems in the RF tuner and IF stages. Thus, the problem should be localized first to the video circuits by monitoring the tuner (Sec. 6-6.2.7) and the IF stages (Sec. 6-7.2.5). If the intermittent condition occurs but the tuner and IF indications are good, the problem is in the video amplifier circuits.

If the intermittent symptom is in the video only, the problem is then likely to be in the video amplifier circuits—probably in the video output transistor Q_2 circuits. This can be confirmed by monitoring the video signal at the base of Q_2 and the cathode of the picture tube. If the intermittent condition occurs without change in the video signals at the picture-tube cathode and Q_2 base, the problem is in the picture tube. Monitor all of the picture-tube voltages. Then try replacement of the picture tube. If the signal at the base of Q_2 remains unchanged but the picture-tube cathode signal is intermittent, the problem is in the video output circuit.

6–9. AGC CIRCUITS

Most solid-state TV sets use a keyed, saturation-type AGC circuit. The RF tuner and IF stage transistors connected to the AGC line are forward-biased at all times. On strong signals, the AGC circuits increase the foward bias, driving the transistors into saturation, thus reducing gain. Under no-signal conditions, the forward bias remains fixed.

Although the AGC bias is a d-c voltage, it is partially developed (or controlled) by bursts of IF signals. A portion of the IF signal is taken from the IF amplifiers and is pulsed, or keyed, at the horizontal sweep frequency rate of 15,750 Hz. The resultant keyed bursts of signal control the amount of d-c voltage produced on the AGC line.

Figure 6–19 is the schematic diagram of a typical solid-state AGC circuit. Transistor Q_1 is an IF amplifier, with its collector tuned to the IF center frequency of 42 MHz by transformer T_1. No d-c voltage as such is supplied to the elements of Q_1. The keying pulses from the horizontal flyback transformer are applied to the collector through diode CR_1. This produces an average collector voltage of about 1 V. When Q_1 is keyed on, the bursts of IF signals pass through T_1 and are rectified by CR_2. A corresponding d-c voltage is developed across C_4 and acts as a bias for AGC amplifier Q_2. Transistor Q_2 is connected as an emitter follower, with the AGC line being

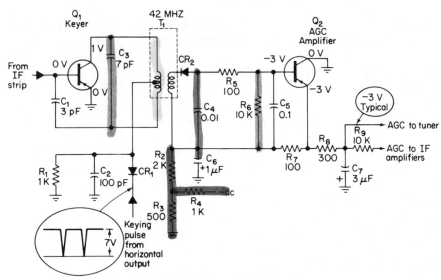

Figure 6–19. Keyed AGC circuits.

returned to the emitter. Variations in IF signal strength cause corresponding variations in Q_2 bias, Q_2 emitter voltage, and the AGC line voltage.

6–9.1. Recommended Troubleshooting Approach

If the AGC circuits are suspected of producing any problems (typically, picture pulling, weak picture, brightness modulation, overloading, picture bending, picture washout, etc.), clamp the AGC line. That is, apply a fixed d-c voltage to the AGC line equal to the normal AGC voltage. If the trouble symptom is removed with a fixed voltage applied, the trouble is likely to be in the AGC circuits.

The next step is to monitor the keying pulse to the AGC circuits and all of the voltages at the transistors. In the circuit of Fig. 6–19, the collector is the only element of Q_1 with a measurable voltage, and this voltage depends on the keying pulse.

A few solid-state AGC circuits are provided with an AGC control that sets the level of AGC action. This control is an internal screwdriver-type adjustment. Always try to correct AGC problems by adjustment before proceeding with troubleshooting, when the AGC circuits are provided with a control.

6–9.2. Typical Troubles

The following paragraphs discuss symptoms that could be caused by defects in the AGC circuits.

6-9.2.1 No Picture and No Sound

When there is no picture or sound with raster present, clamp the AGC line. If the picture and sound are restored, remove the clamp and measure the AGC line voltage. Check the keying pulse. If the keying pulse is present but the AGC line voltage is abnormal (as it must be if the tuner and IF stages are completely cut off) look for shorted or open capacitors, defective diodes, breaks in printed circuit wiring, cold solder joints, or similar conditions. If the keying pulse is absent, check the flyback transformer winding (Sec. 6–2).

6-9.2.2. Poor Picture

When picture quality is poor (weak picture, overloaded picture, pulling, picture washout, or brightness modulation) and the AGC circuits are suspected, follow the same procedure as for a no-picture, no-sound symptom. That is, clamp the AGC line with the correct forward-bias voltage.

If the problem is cleared indicating an AGC circuit defect, check the keying pulse and measure the AGC circuit voltages. This should pinpoint the circuit problem. A possible exception is open capacitors which must be checked by substitution or by shunting with known good capacitors.

6–10. SOUND IF AND AUDIO CIRCUITS

The basic functions of sound IF and audio circuits in solid-state sets are the same as for vacuum-tube sets. The circuits amplify the 4.5-MHz sound carrier, demodulate the FM sound (remove the audio signals), and amplify the audio signals to a level suitable for reproduction on the loudspeaker. The audio portion of the circuits include a volume control and usually a tone control.

Figure 6–20 is the schematic diagram of typical sound IF and audio circuits. Two stages of IF amplification and two stages of audio amplification are used, with all stages in the common-emitter configuration. All stages are forward-biased. The final audio stage is push/pull and therefore operates class B (with slight forward bias) or possibly class AB (with slightly increased forward bias). Unlike the video IF stages, the sound IF stages do not have any AGC circuits, and the forward bias is provided by fixed-resistance networks.

Both sound IF stages Q_1 and Q_2 are neutralized to prevent oscillation. Each stage is tuned at the output by corresponding transformers T_1 and T_2 to 4.5 MHz. The overall bandwidth of the sound IF amplifiers is typically 50 to 60 kHz in solid-state circuits.

There are no test points as such provided on most solid-state sound circuits. However, input from the video circuits is usually applied through a coaxial cable or shielded lead. This cable or lead can be disconnected, if it is required for testing, and can be used as a signal injection point for overall sound tests and alignment. The audio portion of the circuit can be tested by injecting an audio signal at the wiper arm of the volume control. The audio output voltage from the detector is typically less than 1 V.

Note that a ratio detector circuit is used. This is typical. However, some solid-state sets use a discriminator for the sound detector.

6–10.1. Recommended Troubleshooting Approach

The recommended approach for troubleshooting sound stages depends largely on the available test equipment. Ideally, an analyzer-type generator is used. These generators have a 4.5-MHz output that can be modulated (FM) by an audio tone (typically 1 kHz). This composite signal is injected to the sound IF input, and the resultant tone is monitored on the loudspeaker.

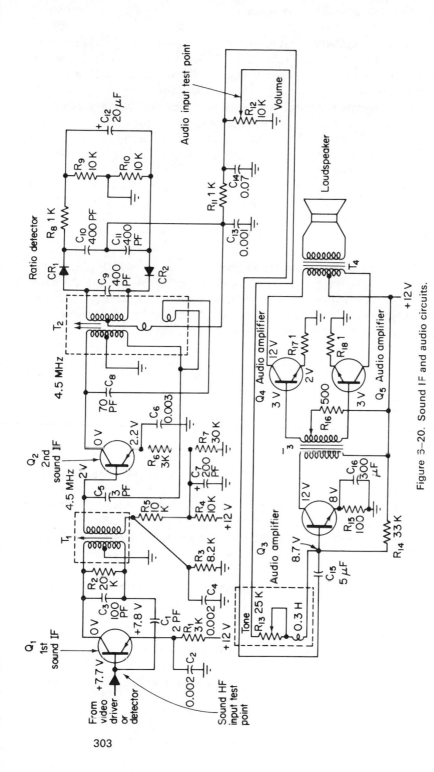

Figure 5-20. Sound IF and audio circuits.

303

The tone can be disabled so that the 4.5-MHz signal is used for alignment of the IF stages. Also, the audio tone can be injected (separately from the 4.5-MHz carrier) into the audio stages (usually at the volume control wiper arm) Thus, the audio stages can be checked separately from the IF stages.

If an analyzer-type generator is not available, the next recommended test setup is a sweep generator with markers and an oscilloscope. The sweep generator signal is injected at the sound IF input, and the signal is monitored at various points throughout the IF stages and at the detector output with an oscilloscope. A demodulator probe is required for the oscilloscope if individual IF stages are to be monitored. A basic low-capacitance probe can be used if the output of the sound detector is monitored. Keep in mind that signals through the sound IF stages of a solid-state TV are quite low (usually in the order of 1 V). Therefore, the oscilloscope must have considerable vertical gain.

If analyzer and sweep generators are not available, use an RF generator to align the sound IF stages, then inject an audio generator signal to the audio stages (volume control wiper). If a trouble such as distortion is definitely isolated to the audio section, a squarewave test can be made. The procedure is similar to that described for video amplifiers in Sec. 6–8.2.6.

6–10.2. Typical Troubles

The following paragraphs discuss symptoms that could be caused by defects in the sound IF and audio stages.

6-10.2.1. No Sound

If the picture is normal but there is no sound, first inject a 4.5-MHz signal at the sound IF amplifier input. Next, inject an audio signal (typically 1 kHz) at the audio amplifier input (volume control). If the audio signal passes but the IF signal is blocked, the problem is in the IF stages or possibly the detector. If the audio signal does not pass, the problem is likely to be in the audio section.

Next inject signals (audio or IF, as applicable) at the base of each transistor and monitor the response on the loudspeaker. If this does not localize the problem immediately, measure the voltage at all transistor elements. This should isolate any problem serious enough to cause a no-sound symptom. A possible exception is an open bypass capacitor. If an emitter bypass capacitor should open, the stage gain (a-c gain) will drop considerably, possibly without seriously affecting the transistor voltages. As usual, a suspected open capacitor can be checked by connecting a known good capacitor in parallel and reapplying power.

6-10.2.2. Poor or Weak Sound

The basic procedure for troubleshooting a poor sound symptom (weak sound, buzzing, distortion, etc.) is the same as for no-sound. That is, the trouble must first be isolated to the IF stages, the audio stages, or the detector and then to a particular stage. This is done by signal injection. Next, the transistor voltages are measured. Generally, a poor sound trouble is more difficult to isolate than a no-sound symptom. For example, the service literature rarely gives the gain per stage of either the sound IF stage or the audio stages.

As a rule of thumb, both the video detector (the input to the sound IF stages) and the audio detector (input to the audio stages) produce about 1 V in normal operation. This output can be as low as 0.5 V but is rarely greater than 2 V. In any event, if a 1 V audio signal is injected at the input of the audio stages (audio detector output or volume control) and the volume control is set to about midrange, there should be a loud tone heard on the loudspeaker. Likewise, a 1 V 4.5-MHz signal (frequency modulated by an audio tone) injected at the sound IF input should also produce a loud signal on the loudspeaker. With the modulated signal applied, it should be possbile to measure about 1 V at the audio detector output. Keep in mind that the ratio detector output is zero when the 4.5-MHz signal is not frequency modulated.

If the trouble appears to be in the sound IF stages, try correcting the problem by alignment. The sound IF stages rarely go out of alignment during normal operation (except in the hands of a do-it-yourselfer). That is, circuit conditions usually do not change so much that the transformers must be drastically retuned. However, aging of components may require periodic peaking of the alignment controls. When an IF transformer must be seriously retuned and it is confirmed that the stages have not been tampered with by unskilled hands, look for a defective component that is changing the resonant frequency of the transformer circuit. Open capacitors are likely suspects.

If the symptom is a weak signal and the IF stages are suspect, measure the gain of both stages. Generally, most of the gain is produced in the second IF stage (Q_2 of Fig. 6–20). Typically, the output from the second IF stage is about 3 to 4 V. If the gain of Q_2 is not at least 10 (usually higher), Q_2 is probably defective. Thus, if the base of Q_2 shows a 0.3 -to 0.4-V signal, the output to the ratio detector should be 3 to 4 V.

If the sound is very low with a very weak background noise, the most likely suspects are leaking capacitors and/or leaking transistors. Of course, this condition should show up as abnormal voltage indications. Very weak sound can be caused by an open emitter (or base) bypass capacitor. This condition will reduce a-c gain but leaves the d-c voltages relatively unaffected.

If the sound is weak, with a loud buzz, the most likely suspect is the ratio detector. In effect, the ratio detector is not rejecting amplitude modulation. Since the sound signal may also contain some 60-Hz buzz from the vertical sync pulses (present at the video detector output), this signal may be amplitude modulating the sound. Normally, the limiting action of Q_2 and the function of the ratio detector will prevent any audio from passing. If the ratio detector is defective, the most likely suspects are diodes CR_1 and CR_2 or stabilizing capacitor C_{12}.

As a general rule with audio circuits, if the signal is weak but the background noise is strong, look for an open condition in the signal path (open coupling capacitors, worn volume controls, etc.). If both the signal and background noise are weak, look for shorted or leaking components (capacitors, transistors, etc.)

When troubleshooting the audio section, *do not remove or disconnect the loudspeaker from the circuit*. If the final output stages of any solid-state audio circuit are operated without a load, the inverse voltage developed across the output transformer can cause the final transistors to break down. If the loudspeaker must be disconnected for any reason, connect a load resistance in its place across the transformer secondary winding. Use a resistance value equal to the loudspeaker impedance (typically, 3.2, 4, 8, or 16 Ω). The load resistance must be capable of handling the full output of the final stages (in watts). If a lower wattage resistor is used, it can burn out, removing the load and causing breakdown of the final transistors.

Index

A

A-C coupling 106
A/D conversion 137
Adders 150
Adjustments, internal 48
AGC circuits 300
Amplifiers 102, 113
Amplifier gain, effects of leakage 91
Amplifier testing 178
AND gates 106
AND gate testing 175
AVC circuit 37

B

BCD–decade conversion 133
BCD–readout conversion 135
BCD signal formats 137
BCD system 100
Bias problems, solid-state oscillators 87
Binary logic 99
Binary numbers 100
Block diagram, example of localizing
    trouble 18
Block diagram, functional 15
Block diagram, servicing 26, 30
Blocking oscillator 89

C

Capacitors in solid-state circuits 93
Capacitor pulse generator 164
Checkout, operational 51
Circuit group 13, 27
Circuit troubles 5
Class A oscillator 88

Class C oscillator 89
Cold solder joints 98
Comparison circuit 152
Component testing 67
Counter readout 131
Counter troubleshooting 194

D

D/A conversion 143
Data recording 11
D-C stability problems 216
Decade circuits 131
Decoders 153
Delay elements 104, 124
Delay FF 122
Delay testing 183
Desoldering tool for ICs 79
Differential amplifier troubleshooting
    226
Digital circuits, introduction to 98
Digital circuit testing 173
Digital component application 131
Digital equipment service literature 183
Digital logic elements 102
Digital measurements 165
Digital test equipment 157
Digital troubleshooting 156
Diodes, testing 66
Divider circuit 131
Dual in-line ICs 83

E

Encode gates 113
Encode gate testing 178

Encoders 153
Encoder, practical 141
Equations, logic 130
Equipment failure 10
Evaluating symptoms 10
EXCLUSIVE OR gates 112
EXCLUSIVE OR testing 176

F

Failure analysis 3, 6
Feedback amplifier troubleshooting 200
Feedback path 34
FFs 115
FF applications 117
FF testing 180
Flat packs, IC 80
Forward bias method of transistor
    testing 60
Four-bit system 138
Full adder 151

G

Gain, transistor 65
Gates 102
Gobbler, solder 74
Go/No-go tests 62
Good input/bad output technique 17

H

Half-adder 150
Half-adder troubleshooting 192
Half-split technique 29
Horizontal oscillator & driver
    troubleshooting 256
H-V supply & horizontal output
    troubleshooting 245

I

IC mounting patterns 80
IC reference designations 127
ICs, soldering 78
IC testing 68
IC troubleshooting sequence 15
IC voltage measurements 69
IF and video detector troubleshooting
    286
IGFETs, handling 67
Industrial equipment troubleshooting
    199
Inspection of equipment 41

Integrator troubleshooting 223
Internal adjustments 48
Inversion 105
Inverters 113
Inverter testing 178
Isolating trouble 4, 25, 32, 35

J

J-K FF 120

L

Logic clip 161
Logic diagrams 185
Logic elements 102
Logic equations 130
Logic probe 159
Logic symbology 98
Logic symbol identification 125
Laboratory equipment troubleshooting
    199
Latching FF 121
Lead bending 84
Leakage effects on amplifier gain 91
Leakage, transistor 65
Linear path 33
Localizing trouble 3
Localizing trouble to a module 12
Locating a specific trouble 40
Location information 129
Low voltage 96
L-V power-supply troubleshooting 238

M

Modules 13
More than one fault 49
Meeting path 33
MV 122
MV testing 181

N

NAND gates 110
NAND gate testing 177
NOR gates 111
NOR gate testing 177

O

One-shot MV 122
Operating sequence 7
Operational amplifier troubleshooting
    205